RIBA Plan of Wor
Contract Administration

The RIBA Plan of Work 2013 Guides

Other titles in the series:

Design Management, by Dale Sinclair

Project Leadership, by Nick Willars

Town Planning, by Ruth Reed

Coming in 2015:

Information Exchanges

Sustainability

Conservation

Health and Safety

Handover Strategy

The RIBA Plan of Work 2013 is endorsed by the following organisations:

Royal Incorporation of Architects in Scotland

Chartered Institute of Architectural Technologists

Royal Society of Architects in Wales

Construction Industry Council

Royal Society of Ulster Architects

RIBA Plan of Work 2013 Guide

Contract Administration

Ian Davies

RIBA Publishing

Published by RIBA Publishing, The Old Post Office, St Nicholas Street, Newcastle upon Tyne NE1 1RH

ISBN 978 1 85946 552 3
Stock code 82652

British Library Cataloguing in Publication Data
A catalogue record for this book is available from the British Library.

Commissioning Editor: Sarah Busby
Series Editor: Dale Sinclair
Project Manager: Alasdair Deas
Design: Kneath Associates
Typesetting: Academic+Technical, Bristol, UK
Printed and bound by CPI Group (UK) Ltd
Cover image: © Iwan Baan, Wakefield Council

Picture credits
The following figures are reproduced with permission: 5.4, 5.8, 5.11, 5.13, 5.14, 5.16, 6.3, 6.4: NBS Contract Administrator

RIBA Publishing is part of RIBA Enterprises Ltd
www.ribaenterprises.com

Contents

Foreword

When I was working in an office many years ago and studying for what is now known as the 'Part III', I searched in vain for some book which would give me clear and authoritative answers to my questions about contract administration.

In those days it was rare to see a building contract, much less know anything about it. The instruction on building contracts that I received at university lasted exactly an hour and a half late one Friday afternoon. We were given copies of the form of contract and we were taken through it at a rapid pace. I still have the notes I took on that day. Under 'extension of time' I had written: 'It is the architect's job to give extensions of time'. The rest was no better. I never found the elusive book, and even many years later, when I was actually administering contracts for a local authority, I was still feeling my way.

In this book, contract administration is clearly and precisely explained and many common questions are answered in the process. How I wish I had had this book, not only when I was working for the Part III, but also years later, when I was regularly faced with administering building contracts that I barely understood. I thoroughly recommend it to architects young and old.

Dr David Chappell
Architect, Building Contract Consultant, Adjudicator and
Professor in Architectural Practice and Law.
Author of over 60 books for the construction industry.

Series editor's foreword

The RIBA Plan of Work 2013 was developed in response to the needs of an industry adjusting to emerging digital design processes, disruptive technologies and new procurement models, as well as other drivers. A core challenge is to communicate the thinking behind the new RIBA Plan in greater detail. This process is made more complex because the RIBA Plan of Work has existed for 50 years and is embodied within the psyche and working practices of everyone involved in the built environment sector. Its simplicity has allowed it to be interpreted and used in many ways, underpinning the need to explain the content of the Plan's first significant edit. By relating the Plan to a number of commonly encountered topics, the *RIBA Plan of Work 2013 Guides* series forms a core element of the communication strategy and I am delighted to be acting as the series editor.

The first strategic shift in the RIBA Plan of Work 2013 was to acknowledge a change from the tasks of the design team to those of the project team: the client, design team and contractor. Stages 0 and 7 are part of this shift, acknowledging that buildings are used by clients, or their clients, and, more importantly, recognising the paradigm shift from designing for construction towards the use of high-quality design information to help facilitate better whole-life outcomes.

New procurement strategies focused around assembling the right project team are the beginnings of significant adjustments in the way that buildings will be briefed, designed, constructed, operated and used. Design teams are harnessing new digital design technologies (commonly bundled under the BIM wrapper), linking geometric information to new engineering analysis software to create a generation of buildings that would not previously have been possible. At the same time, coordination processes and environmental credentials are being improved. A core focus is the progressive fixity of high-quality information – for the first time, the right information at the right time, clearly defining who does what, when.

The RIBA Plan of Work 2013 aims to raise the knowledge bar on many subjects, including sustainability, Information Exchanges and health and safety. The *RIBA Plan of Work 2013 Guides* are crucial tools in disseminating and explaining how these themes are fully addressed and how the new Plan can be harnessed to achieve the new goals and objectives of our clients.

Dale Sinclair
November 2014

Acknowledgements and dedication

To my wife Lynda for her support and patience, and to the many clients, contractors and colleagues I have known over the years who have given me a valuable insight into the procedures and the problems associated with managing building contracts.

I would like to thank Dale Sinclair, who chaired the group which developed the RIBA Plan of Work 2013, for his help and comments in the preparation of this book, and Sarah Busby for her editorial guidance.

My thanks also go to Jeremy Baxter at Gallaghers for advice on insurances, to Rob Ellison at NBS for arranging for me to use a copy of NBS Contract Administrator and to John Wevill of Clarkslegal LLP for his valuable help in reviewing the book.

About the author

Ian has 40 years of experience as a qualified architect and has run and managed all types of building contracts, including traditional and design and build, on both large and small projects. Having been director of a small practice for the past 20 years he has experience of all aspects of practice and a keen understanding of the pressures of the contract administrator role. He now works as a consultant architect and party wall surveyor. He previously worked for Sheppard Robson, The Inskip Partnership and DLP Design Ltd among others and has been involved in a variety of sectors, with a particular focus on education, housing and leisure.

Ian was a key member of the RIBA working group that developed the RIBA Plan of Work 2013. He is also a member of the RIBA Planning Group and remains a corresponding member of the RIBA Small Practice Committee. He was previously involved in the RIBA Procurement Reform Group. He acted as a consultant reviewer of the RIBA Agreements 2010 and for *Assembling a Collaborative Project Team* in 2013.

About the series editor

Dale Sinclair is Director of Technical Practice for AECOM's architecture team in EMEA. He is an architect and was previously a director at Dyer and an associate director at BDP. He has taught at Aberdeen University and the Mackintosh School of Architecture and regularly lectures on BIM, design management and the RIBA Plan of Work 2013. He is passionate about developing new design processes that can harness digital technologies, manage the iterative design process and improve design outcomes.

He is currently the RIBA Vice President, Practice and Profession, a trustee of the RIBA Board, a UK board member of BuildingSMART and a member of various CIC working groups. He was the editor of the *BIM Overlay to the Outline Plan of Work 2007*, edited the RIBA Plan of Work 2013 and was author of its supporting tools and guidance publications: *Guide to Using the RIBA Plan of Work 2013* and *Assembling a Collaborative Project Team*.

Introduction

Overview

When a member of the public looks at a building they see the end product, the result of the design and construction process, but not the complex procedures and actions required to achieve that product. Each project is unique, with many unknown issues at the outset, and it is the successful interaction of all the parties under the control of the project leaders which enables the success of the project. For this you need a plan.

The RIBA Plan of Work 2013 organises the process of briefing, designing, constructing, maintaining, operating and using buildings into a number of key stages. The purpose of this guide is to provide an explanation of the practical matters an architect or other building professional should consider in the administration of a building contract through each of the Plan of Work stages.

In some respects the contract administrator role has changed little over many years, but the introduction of the RIBA Plan of Work 2013 – with its more complex procurement routes and project management role – provides the opportunity to revisit the tasks and procedures involved at each stage, particularly for the small practitioner on smaller projects.

This guide examines the contract administration issues typically encountered at each stage of the RIBA Plan of Work 2013 and how the contract administrator can control the process as the project progresses. It also considers working methods and provides checklists, tools and techniques to make the procedures more efficient and thus more cost effective.

What is the role of the contract administrator and the project team?

The contract administrator is responsible for the administration of the Building Contract, including the issuing of additional instructions and the various certificates required to allow for handover and occupation of the building until all the defects have been rectified and the contract concluded.

As outlined in the publication *Assembling a Collaborative Project Team*, every project, whether large or small, requires a number of roles to be undertaken. On larger projects it is likely that all the core project roles will be required – with the need for them varying from stage to stage – and it is likely that these roles will be undertaken by different parties.

On smaller projects the opposite is true as roles are typically combined, with the appointed architect carrying out the project lead, lead designer and contract administrator roles.

That said, on both large and small projects there are some stages when the contract administrator has only a minor role, with their main input being at the appropriate procurement stage and Stages 4 to 6.

Although traditionally it is the architect who has undertaken the contract administrator role, as projects become ever more complex this role is increasingly being undertaken by other specialists within the project team. Knowledge of what is involved in this role, as presented in this guide, would also be useful to others, such as:

---

contractors who have contract management personnel and employ contracts managers who wish to better understand the contract administrator's role and procedures in contract administration

---

project managers or cost consultant practices who would like to understand more about this subject and how their own services relate to it

---

any member of the project and/or design team who wishes to have a better understanding of how their inputs fit into the construction process

---

students, as a textbook and a good Part 3 guidance document.

Throughout the guide reference is made to the role of the contract administrator irrespective of which discipline is carrying out the role. However, the detailed requirements of the role may vary between projects, depending on the procurement route selected, etc. These differences have been highlighted in the text.

How is procurement identified in the RIBA Plan of Work 2013?

Within the RIBA Plan of Work 2013 procurement is set out as a generic task bar. This is because tendering activities do not have a specific place in the chronology of a project and will be dependent on the form of building contract and procurement route being used. To allow for different forms of procurement, users of the RIBA Plan of Work 2013 can create their own bespoke Plan of Work to suit their selected procurement route.

Once a bespoke Plan of Work has been created for the preferred procurement route, the tasks will vary according to the selected route and the tasks and procedures will be addressed at the appropriate stage.

Bespoke Plans of Work are currently available at **www.ribaplanofwork.com** for five forms of procurement.

Within a bespoke Plan of Work the variable procurement task bar will fix the support tasks for the contract administrator relating to

Tasks ▼ / Stages ▶	0 Strategic Definition	1 Preparation and Brief	2 Concept Design	3 Developed Design
Procurement	Initial considerations for assembling the project team.	Prepare **Project Roles Table** and **Contractual Tree** and continue assembling the project team.	← The procurement strategy does not fundamentally alter the progression of the design or the level of detail prepared at	a given stage. However, Information Exchanges will vary depending on the selected procurement route and Building Contract. A bespoke RIBA Plan of Work

Figure 1 The procurement task bar from the RIBA Plan of Work 2013

When should you go to tender?

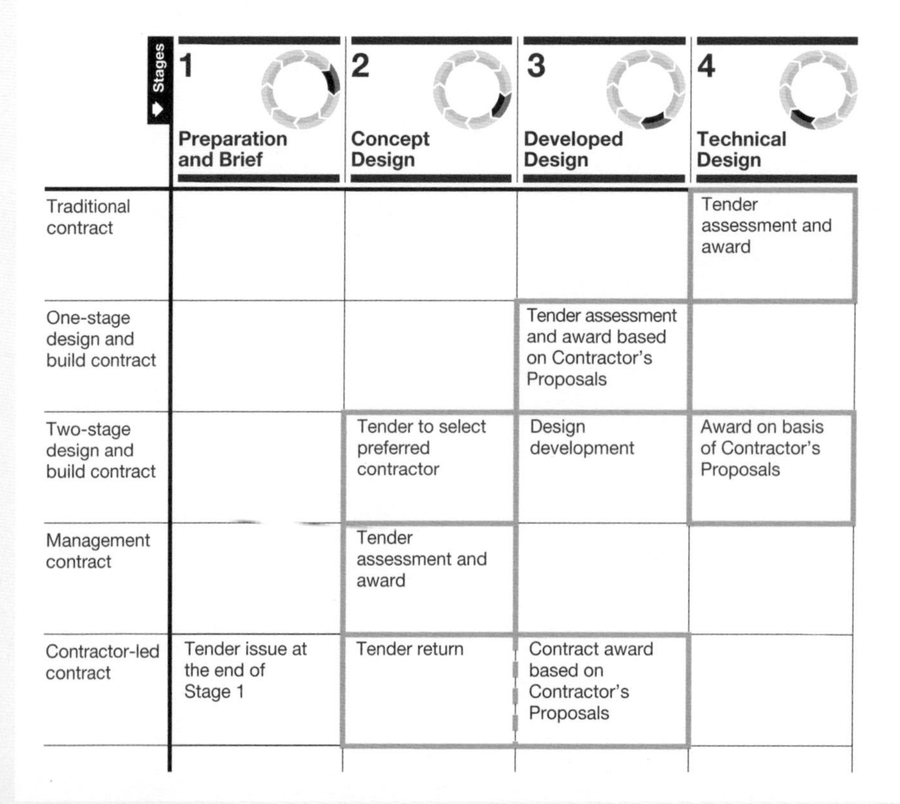

Stages ▶	1 Preparation and Brief	2 Concept Design	3 Developed Design	4 Technical Design
Traditional contract				Tender assessment and award
One-stage design and build contract			Tender assessment and award based on Contractor's Proposals	
Two-stage design and build contract		Tender to select preferred contractor	Design development	Award on basis of Contractor's Proposals
Management contract		Tender assessment and award		
Contractor-led contract	Tender issue at the end of Stage 1	Tender return	Contract award based on Contractor's Proposals	

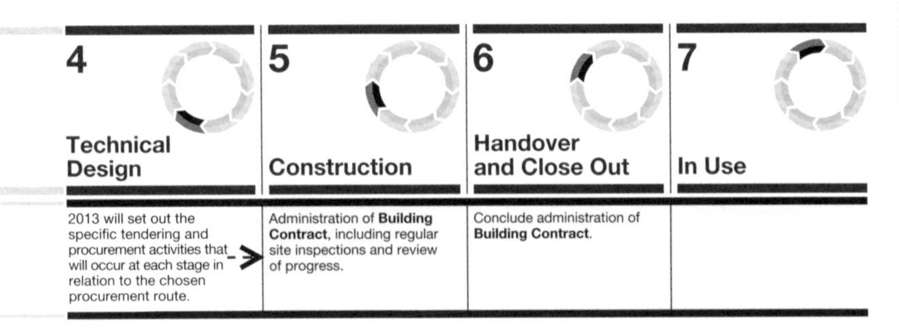

the issue of tender documents, receipt of tenders and the awarding of the Building Contract. The key outputs will include finalising tender documents, which will incorporate drawings, specifications, schedules of work, health and safety information, terms of bonds and warranties and other tender information.

By generating these bespoke plans it can be noted that tendering procedures and contract award occur at different stages for the alternative forms of procurement and this will be addressed at the appropriate Plan of Work stage:

- traditional at Stage 4
- one-stage design and build at Stage 3
- two-stage design and build at Stages 2 and 4
- management contract at Stage 2
- contractor-led contract at Stages 1, 2 and 3.

How to use this guide

Although this guide to contract administration can be used on its own, it can also be used in conjunction with the other handbooks in the RIBA Plan of Work 2013 series as well as other RIBA publications, such as the *RIBA Job Book*.

The structure of the guide follows the Plan of Work stages, from 0 to 7. Each chapter starts with an overview followed by details of the tasks and procedures the contract administrator would be expected to undertake for that stage. This is accompanied by checklists and typical examples which can be adapted to suit the nature and size of the project.

Tendering procedures are covered in Stage 4 although the timing of tendering activities will depend on the type of procurement route selected. The timing of the tender process has been highlighted at the stage appropriate for the procurement route.

The RIBA Plan of Work 2013 is available online at **www.ribaplanofwork.com**. As well as explaining in more detail the concept and content of each stage, this has been developed as a flexible tool to enable the creation of a bespoke practice- or project-specific Plan of Work containing the relevant procurement (tendering), programme and town planning activities. It can be tailored to accommodate specific project and client requirements.

With each procurement route there are a number of building contract forms that can be used: RIBA, JCT, NEC3, FIDIC and PPC2000. The choice of contract form is often dictated by the client, whether private or public sector, and the contract administrator will require an in-depth knowledge of the particular building contract being used. This guide gives an overview of the different forms, identifies where there are differences between them and provide links to where further details can be obtained.

When using JCT contracts it is suggested that this book is used in association with the *NBS Contract Administrator* software available from **www.thenbs.com**. *NBS Contract Administrator* provides an efficient platform for managing JCT/SBCC contracts and contains electronic copies of the official RIBA contract administration forms for use with the JCT standard, intermediate, minor works, and design and build contracts. These forms cover all the main activities, such as instructions, extension of time, interim/final payments and Practical Completion – the software filters the right forms for the particular contract at the appropriate time.

Ian Davies
November 2014

Using this series

For ease of reference each book in this series is broken down into chapters that map on to the stages of the Plan of Work. So, for instance, the first chapter covers the tasks and considerations around contract administration at Stage 0.

We have also included several in-text features to enhance your understanding of the topic. The following key will explain what each icon means and why each feature is useful to you:

The 'Example' feature explores an example from practice, either real or theoretical

The 'Tools and Templates' feature outlines standard tools, letters and forms and how to use them in practice

The 'Signpost' feature introduces you to further sources of trusted information from books, websites and regulations

The 'Definition' feature explains key terms in this topic area in more detail

The 'Hints and Tips' feature dispenses pragmatic advice and highlights common problems and solutions

The 'Small Project Observation' feature highlights useful variations in approach and outcome for smaller projects

RIBA

The **RIBA Plan of Work 2013** organises the process of briefing, designing, constructing, maintaining, operating and using building projects into a number of key stages. The content of stages may vary or overlap to suit specific project requirements.

RIBA Plan of Work 2013

Tasks / Stages	0 Strategic Definition	1 Preparation and Brief	2 Concept Design	3 Developed Design
Core Objectives	Identify client's **Business Case** and **Strategic Brief** and other core project requirements.	Develop **Project Objectives**, including **Quality Objectives** and **Project Outcomes**, **Sustainability Aspirations**, **Project Budget**, other parameters or constraints and develop **Initial Project Brief**. Undertake **Feasibility Studies** and review of **Site Information**.	Prepare **Concept Design**, including outline proposals for structural design, building services systems, outline specifications and preliminary **Cost Information** along with relevant **Project Strategies** in accordance with **Design Programme**. Agree alterations to brief and issue **Final Project Brief**.	Prepare **Developed Design**, including coordinated and updated proposals for structural design, building services systems, outline specifications, **Cost Information** and **Project Strategies** in accordance with **Design Programme**.
Procurement *Variable task bar	Initial considerations for assembling the project team.	Prepare **Project Roles Table** and **Contractual Tree** and continue assembling the project team.	← The procurement strategy does not fundamentally alter the progression of the design or the level of detail prepared at	a given stage. However, **Information Exchanges** will vary depending on the selected procurement route and **Building Contract**. A bespoke **RIBA Plan of Work**
Programme *Variable task bar	Establish **Project Programme**.	Review **Project Programme**.	Review **Project Programme**.	← The procurement route may dictate the **Project Programme** and result in certain stages overlapping
(Town) Planning *Variable task bar	Pre-application discussions.	Pre-application discussions.	← Planning applications are typically made using the Stage 3 output.	A bespoke **RIBA Plan of Work 2013** will identify when the
Suggested Key Support Tasks	Review **Feedback** from previous projects.	Prepare **Handover Strategy** and **Risk Assessments**. Agree **Schedule of Services**, **Design Responsibility Matrix** and **Information Exchanges** and prepare **Project Execution Plan** including **Technology** and **Communication Strategies** and consideration of **Common Standards** to be used.	Prepare **Sustainability Strategy**, **Maintenance and Operational Strategy** and review **Handover Strategy** and **Risk Assessments**. Undertake third party consultations as required and any **Research and Development** aspects. Review and update **Project Execution Plan**. Consider **Construction Strategy**, including offsite fabrication, and develop **Health and Safety Strategy**.	Review and update **Sustainability**, **Maintenance and Operational** and **Handover Strategies** and **Risk Assessments**. Undertake third party consultations as required and conclude **Research and Development** aspects. Review and update **Project Execution Plan**, including **Change Control Procedures**. Review and update **Construction** and **Health and Safety Strategies**.
Sustainability Checkpoints	**Sustainability Checkpoint – 0**	**Sustainability Checkpoint – 1**	**Sustainability Checkpoint – 2**	**Sustainability Checkpoint – 3**
Information Exchanges (at stage completion)	**Strategic Brief**.	**Initial Project Brief**.	**Concept Design** including outline structural and building services design, associated **Project Strategies**, preliminary **Cost Information** and **Final Project Brief**.	**Developed Design**, including the coordinated architectural, structural and building services design and updated **Cost Information**.
UK Government Information Exchanges	Not required.	Required.	Required.	Required.

***Variable task bar** – in creating a bespoke project or practice specific RIBA Plan of Work 2013 via www.ribaplanofwork.com a specific bar is selected from a number of options.

The **RIBA Plan of Work 2013** should be used solely as guidance for the preparation of detailed professional services contracts and building contracts.

www.ribaplanofwork.com

	4 Technical Design	**5** Construction	**6** Handover and Close Out	**7** In Use
	Prepare **Technical Design** in accordance with **Design Responsibility Matrix** and **Project Strategies** to include all architectural, structural and building services information, specialist subcontractor design and specifications, in accordance with **Design Programme**.	Offsite manufacturing and onsite **Construction** in accordance with **Construction Programme** and resolution of **Design Queries** from site as they arise.	Handover of building and conclusion of **Building Contract**.	Undertake **In Use** services in accordance with **Schedule of Services**.
	2013 will set out the specific tendering and procurement activities that will occur at each stage in relation to the chosen procurement route. ->	Administration of **Building Contract**, including regular site inspections and review of progress.	Conclude administration of **Building Contract**.	
	or being undertaken concurrently. A bespoke **RIBA Plan of Work 2013** will clarify the stage overlaps.	The **Project Programme** will set out the specific stage dates and detailed programme durations. ->		
	planning application is to be made. ->			
	Review and update **Sustainability, Maintenance and Operational** and **Handover Strategies** and **Risk Assessments**. Prepare and submit Building Regulations submission and any other third party submissions requiring consent. Review and update **Project Execution Plan**. Review **Construction Strategy**, including sequencing, and update **Health and Safety Strategy**.	Review and update **Sustainability Strategy** and implement **Handover Strategy**, including agreement of information required for commissioning, training, handover, asset management, future monitoring and maintenance and ongoing compilation of **'As-constructed' Information**. Update **Construction** and **Health and Safety Strategies**.	Carry out activities listed in **Handover Strategy** including **Feedback** for use during the future life of the building or on future projects. Updating of **Project Information** as required.	Conclude activities listed in **Handover Strategy** including **Post-occupancy Evaluation**, review of **Project Performance**, **Project Outcomes** and **Research and Development** aspects. Updating of **Project Information**, as required, in response to ongoing client **Feedback** until the end of the building's life.
	Sustainability Checkpoint — 4	**Sustainability Checkpoint — 5**	**Sustainability Checkpoint — 6**	**Sustainability Checkpoint — 7**
	Completed **Technical Design** of the project.	**'As-constructed' Information**.	Updated **'As-constructed' Information**.	**'As-constructed' Information** updated in response to ongoing client **Feedback** and maintenance or operational developments.
	Not required.	Not required.	Required.	As required.

Stage 0

Strategic Definition

Chapter overview

Stage 0 is when the client's Business Case is clarified and the Strategic Brief agreed and undertaken, with fundamental questions being asked about the project. The contract administrator will not be involved or necessarily appointed at this stage, but strategic decisions on assembling the project team and forms of procurement and building contracts will impact on the role at subsequent stages. This chapter considers how the contract administrator role interfaces with other roles at this stage and considers how advice and decisions made at this stage will affect the role in subsequent stages. The **key coverage in this chapter is as follows**:

- Setting up the project
- An overview of the role of the contract administrator for alternative procurement routes
- Feedback from Stage 7

Introduction

At Stage 0 the project is strategically appraised and defined before the Initial Project Brief is created at Stage 1. Initial considerations will be made regarding assembling the project team and also on the likely form of procurement to be used, depending on the type of project or an experienced client's preferences.

Small projects at Stage 0

On a small project, Stage 0 may simply form part of the initial meeting with the client to consider the client's brief before securing the commission and so will effectively be concurrent with Stage 1. On a larger project, it would be a distinct stage.

It should be remembered, however, that although Stage 0 is one of the core stages, the scope will vary. Therefore, a time charge fee is appropriate for this stage. On larger, more complex projects, it should be seen as a distinct work stage commanding an additional fee.

What are the Core Objectives of this stage?

The Core Objectives of the RIBA Plan of Work 2013 at Stage 0 are:

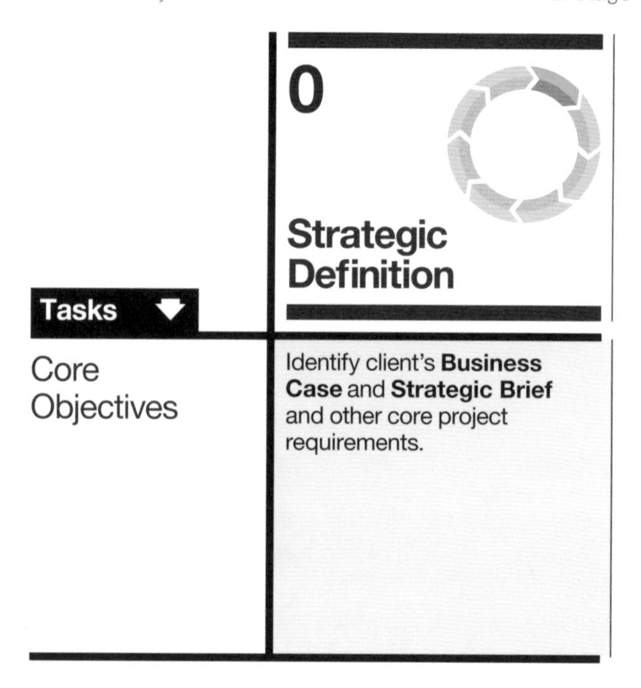

The Core Objectives at Stage 0 relate to identifying the client's Business Case and the preparation of the Strategic Brief, together with identifying other core project requirements.

What supporting tasks should be undertaken during Stage 0?

The Suggested Key Support Tasks noted in the RIBA Plan of Work 2013 have been devised to support the Core Objectives and to ensure that the documentation required to proceed to the next stage has been prepared.

The support tasks during this stage are focused on reviewing the feedback from previous projects, which will have been fed back to the design team by the contract administrator at Stage 7.

What level of service is involved at this stage?

Stage 0 is the initial stage within the RIBA Plan of Work 2013. The level of service can range from minimal on a small project through to a fully defined level of service on a larger project, which would include:

- receiving the client's instructions and information about the project
- assisting in defining the client's strategic requirements and preparing the Business Case and the Strategic Brief
- reviewing alternative project team options with the client
- contributing to the Project Programme and assembling the project team.

What is the project team at this stage?

This stage may not involve the appointment of many members of the design team, but it is likely to require the client, the client adviser, the project lead and possibly the architect (although on larger projects no designers may be appointed). When selecting the right team it is important to ask the client the right questions to properly define the scope of the project and to agree who is to lead the team, whether it is the client, project manager, architect or contractor. Other members of the project team will

Putting together the project team

Assembling a Collaborative Project Team by Dale Sinclair (RIBA Publishing, 2013) gives further guidance on how project teams are procured and assembled.

be appointed at the appropriate stages and the timing of the contractor's involvement can vary according to the preferred procurement route.

Who carries out the contract administrator role?

Although the contract administrator is unlikely to have been officially appointed at this stage, the details of the role will be set out within the professional services contract to be agreed with the client. On traditional appointments, and particularly on small projects such as domestic, the contract administrator role will typically be combined with that of the project lead, lead designer and, possibly, other roles. However, on larger, more complex projects, the contract administrator could be a separate appointment or carried out by the cost consultant or architect. Alternatively it could be the project manager, who may be the main point of contact with the client. All these options need to be addressed at this early stage.

Until it is known at what stage the project is likely to go out to tender, one of the appointed parties may assume the contract administrator role during Stage 0 and Stage 1 in terms of making decisions on procurement and forms of building contract. In such an instance where the 'actual' contract administrator is not giving the initial advice, there is a risk that the client may not be advised appropriately unless the person assuming that role has the relevant knowledge and experience.

What is the role of the contract administrator for various procurement options?

After defining the project team strategically at Stage 0, the assembling of the team continues incrementally at Stage 1 and beyond, leading ultimately to the inclusion of the contractor. However, it is at Stage 0 that the likely role of the contract administrator will be defined, depending on the preferred form of procurement.

Forms of procurement fall into three main categories: traditional, design and build, and management forms. These are addressed in more detail in Stage 1 (pages 30–43). However, when considered strategically, although there are core tasks for the contract administrator applicable to all forms, the role will vary slightly according to the route selected.

Traditional procurement route

On traditional contracts, the contract administrator role will involve carrying out the following core tasks:

- advising the client on methods of procurement, tendering and the appointment of the main contractor
- issuing, receiving and analysing tenders
- preparing contract documents
- administering the terms of the Building Contract
- issuing further information and instructions
- arranging and chairing regular meetings between the contractor, the design team and the client
- managing Change Control Procedures
- certifying payments as required by the Building Contract
- considering claims
- preparing defects lists
- certifying Practical Completion
- actions during the rectification period
- issuing the final certificate
- arranging for the preparation of 'As-constructed' Information and operating and maintenance manuals.

Design and build procurement route

Following a design and build procurement route the project team could work for the client or the contractor, or even both, albeit sequentially rather than simultaneously. If employed by the contractor, the project team would have no function in connection with managing the terms of the Building Contract itself as that is a client appointment. If employed by the client, the project team would advise on tendering procedures and contract matters and, assuming that they are not to be novated to the contractor, one of the team members could act as the employer's agent under the Building Contract.

The employer's agent will normally take on the role of contract administrator and is likely to be either the project lead (often the architect if they have not been novated) or the cost consultant. However, the role can be carried out by someone from the client organisation, such as an in-house project manager, or by an independent practice appointed by the client.

In addition to their contract administrator role, the employer's agent will also carry out other tasks prior to the letting of a contract, such as preparing the Employer's Requirements to be issued with the tender documents, issuing, receiving and analysing the tenders and preparing contract documents.

After the Building Contract has been awarded the contract administrator will undertake the following in addition to the core tasks:

- coordinating the review of detailed drawing and specification information prepared by the contractor
- considering items submitted by the contractor for approval, as required by the Employer's Requirements.

Management contracting procurement routes

Management forms of contracting generally fall into two main categories: management contracting and construction management, neither route being suitable for inexperienced clients and in both instances the contractor is appointed at an early stage.

In such cases the client will appoint someone to be the project manager, who is either an in-house employee or a consultant, to look after the client's interests at all stages. Although the project manager could undertake the contract administrator role, the contract administrator may be a separate individual, often the architect, who would work closely with the project manager throughout the Project Programme period.

Project manager

Further information about the role of the project manager can be found in *The RIBA Job Book*, ninth edition (RIBA Publishing, 2013).

Construction management

On construction management contracts it is the construction manager who will undertake elements of the contract administrator role. In this case the construction manager is employed by the client to organise and manage the trades contractors, who are all employed directly by the client. In practice the construction manager may not be brought in until a late stage during the pre-construction period.

Although the construction manager's remit may include other roles, during the pre-construction stage the construction manager's role as contract administrator will include reviewing and coordinating information prepared by the trades contractors and programme management in addition to the core tasks.

Management contracts

In a management contract either the architect, project manager or a separate contract administrator working with the project manager will undertake management of the Building Contract. The tasks are similar to other procurement routes and, in addition to the core tasks, will include tendering individual trades packages and reviewing and coordinating information prepared by the trades contractors.

What about Feedback from Stage 7 – In Use?

When developing the Strategic Brief one of the Suggested Key Support Tasks is to review Feedback from previous projects so this can be fed into the briefing process for the new project. This feedback could be from other similar but unconnected projects, but the most useful feedback would be from previous work on the same or similar buildings for the same client.

Feedback from Stage 7 feeds back to Stage 0, where data from an existing building can be used to inform the briefing process for the refurbishment or alteration of the building. This is of particular significance where Building Information Modelling (BIM) has been used previously and the updated model would be used to inform the design of the future works.

A post-project review in Stage 7 will have highlighted the major differences between original expectations and actual outcomes and how occupiers use the building. Information from the contract administrator during the construction phase of a previous project would have been fed into the handover manuals to inform the building's users. Also, during the In Use stage (Stage 7), facilities managers and maintenance staff may have produced feedback on whether the building is too complicated to use easily or difficult to maintain. Both these sources of information will benefit the briefing for the next project as well as the subsequent design and construction process.

Chapter summary 0

Although the contract administrator will not be involved or necessarily appointed at this stage, this chapter has set out, from a contract administrator's viewpoint, the initial considerations to be made when assembling the project team and deciding on the form of procurement to be used.

The role and scope of work of the contract administrator will be affected by the form of procurement selected. The contract administrator's input to the Initial Project Brief will be developed as a continuous process though Stage 1 and into Stage 2.

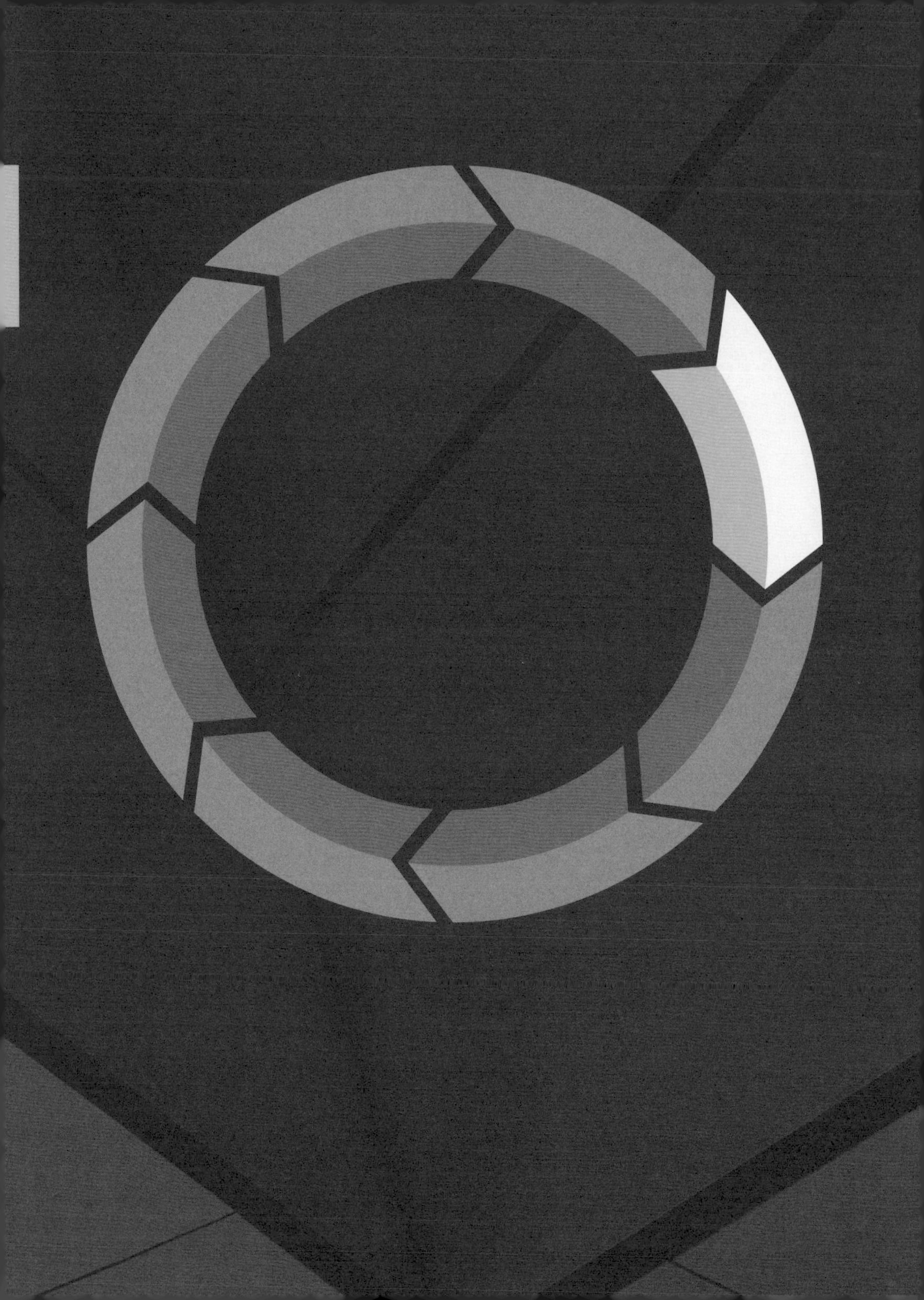

Stage 1

Preparation and Brief

Chapter overview

Although technically there may be little or no involvement for the role of the contract administrator during Stage 1, this chapter considers what questions the project lead should ask or consider from a contract administration perspective when setting up the project.

When assembling the project team the Project Roles Table and the Contractual Tree will be prepared and the Schedules of Services agreed.

This chapter considers alternative forms of procurement and appropriate forms of contract, to enable the project lead to advise on the best way forward to meet the client's needs.

The key coverage in this chapter is as follows:

What is the project team?

What is procurement?

Types of procurement

Selecting the procurement method

Forms of building contract

Tendering at Stage 1

Introduction

Several significant and parallel activities are carried out during Stage 1 to develop the Initial Project Brief and any related Feasibility Studies, together with continuing to assemble the project team and define each party's roles and responsibilities and the Information Exchanges.

The preparation of the Initial Project Brief is the most important task undertaken during Stage 1, together with a project Risk Assessment to establish the risks that need to be managed and monitored. The development of the procurement strategy and the Project Programme are part of this early risk analysis. For Stage 2 to commence in earnest, it is essential that the team is properly assembled by the end of Stage 1.

Although the actual contract administrator may not have been appointed at this point, the role changes according to the size and complexity of the project, from acting as a single point of contact on small projects to being a specialist on larger projects. The main items which will influence how the contract administration is undertaken will be the form of procurement and the form of contract adopted.

Preparation of the Project Roles Table and Contractual Tree will highlight issues for the client that will impact on the contract administration at later stages. The main areas on which the contract administrator will be required to advise the client are the most appropriate form of procurement and the relevant forms of building contract.

The choice of an appropriate procurement method and form of contract will be influenced by external factors. When selecting these, the choices should never be made on an arbitrary basis, but always after a careful analysis of the situation. Whichever procurement route chosen, it is important that the correct form of building contract is used. The options are increasing, particularly for overseas contracts.

What are the Core Objectives of this stage?

The Core Objectives of the RIBA Plan of Work 2013 at Stage 1 are:

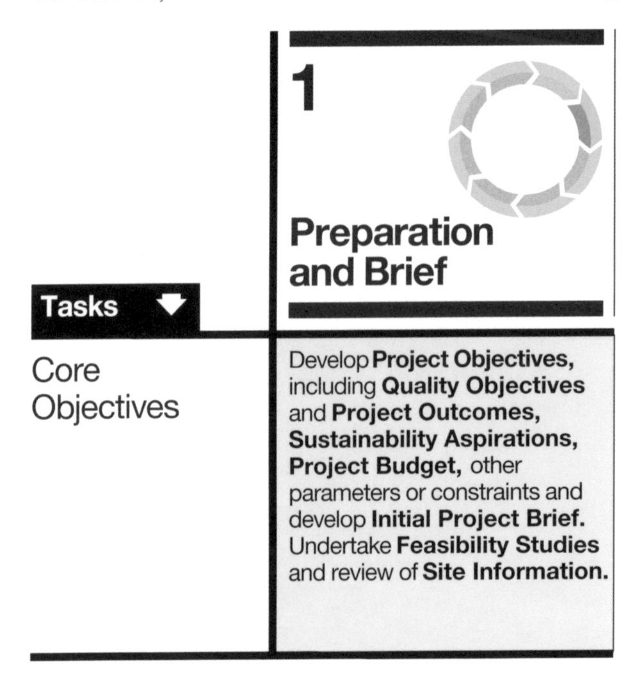

The Core Objectives at Stage 1 revolve around developing the Project Objectives and the Initial Project Brief as well as undertaking any Feasibility Studies. The preparation of the Initial Project Brief is the most important task undertaken during Stage 1 and the time required will very much depend on the complexity of the project.

What supporting tasks should be undertaken during Stage 1?

The Suggested Key Support Tasks noted in the RIBA Plan of Work 2013 have been devised to support the Core Objectives and to ensure that the documentation required to conclude the Building Contract has been prepared.

At this stage the suggested tasks are to 'Prepare Handover Strategy and Risk Assessments' and 'Agree Schedule of Services, Design Responsibility Matrix and Information Exchanges and prepare Project Execution Plan including Technology and Communication Strategies and consideration of Common Standards to be used'.

The support tasks during this stage are focused on ensuring that the project team is properly assembled and that the project Handover Strategy and any post-occupancy services which may be required are considered.

What is the project team?

Together with the client, the project team comprises the design team and the contractor. How the client, the design team and the contractor are connected will depend on the procurement route and how the contractor is introduced to other members of the design team. In many cases the client may be the end user, but it could also be, say, a board of governors, a local authority, etc.

What are the first steps?

While Stage 2 is the considered to be the crucial stage of the project and some might want to start it immediately, the efficient use of Stage 1 can facilitate a more productive Stage 2. One of the first steps is to assemble the Project Roles Table. This requires consideration of the roles that are required for each stage of the project. *Assembling a Collaborative Project Team* splits these roles into the core roles and possible additional roles. On most projects all of the core roles will be required. The need for additional roles will depend on the complexity of the project and the particular issues that have to be addressed.

Having assembled the project team the subsequent decisions to be made by the contract administrator will be the choice of procurement route and the form of building contract to be used.

What is procurement?

Procurement is a generic term embracing all those activities undertaken by a client seeking to bring about the design and construction of a new-build project or the refurbishment of an existing building. Variously referred to as a route, path or system, procurement is initiated by weighing up the benefits, risks and financial constraints that affect the project and which eventually will be reflected in the choice of contractual arrangements. In every project the concerns of the client will primarily focus on time, cost, and performance or quality, relative to the design and the construction of the building.

It should be noted that although the procurement strategy relates to the appointment of the project team, this guide deals with the procurement route leading to the appointment of the main contractor.

A decision on the selection of a procurement method will have resulted from many interrelated factors including the scope and nature of the work, the level of risk, who employs the design team, design responsibility, coordination and the cost basis.

What is the procurement strategy?

The procurement strategy identifies the optimum way of meeting the Project Outcomes, while ensuring value for money and taking into account the risks and constraints. It will involve decisions about the funding mechanism and asset ownership of the project. The aim of the procurement strategy is to achieve the optimum balance of risk, control and funding.

What types of procurement are there?

There is a wide variety of procurement routes – which option is best for a particular project will depend on how the contract is to be managed. These routes can be grouped loosely into three types: traditional, design and build, and management forms.

There are essentially three basic elements that influence the selection of the procurement method: time, quality and cost. However, only two of these are present in any one of the routes. In simple terms, these can be separated out as follows:

traditional	=	**quality** and **cost**	(at the expense of time)
design and build	=	**cost** and **time**	(at the expense of quality)
management	=	**time** and **quality**	(at the expense of cost)

In addition to these basic routes, a development in procurement that has become more commonplace is 'collaborative procurement' or 'partnering'. The intention is that the parties to a project have the mutual objectives of delivering the project on time, to budget and to quality, and is about working as a team, regardless of organisation or discipline, to meet the client's needs. It is most commonly used on large, long-term or high-risk projects.

As well as the three basic elements, consideration also needs to be given to the following factors when selecting the type of procurement method:

- client's knowledge and expertise
- complexity and size of project
- cost: certainty in fluctuations, lowest capital expenditure, best value
- likelihood of design changes
- likelihood of changes during construction
- programme timescale and need for speed: earliest start on site, certainty on contract duration, shortest possible duration
- quality of finished building and level of detail: highest with minimum maintenance, sensitive design with client control, detail not critical contractor design
- responsibility for coordination of work on site
- level of risk to be taken by each party.

What is traditional procurement?

Traditional procurement is the most common form of procurement: it has been used successfully for many years on all sizes and types of building contract. It gives a clear distinction between the design and construction stages, whereby the design team designs and details the works and the contractor builds to the details that the design team produces. The

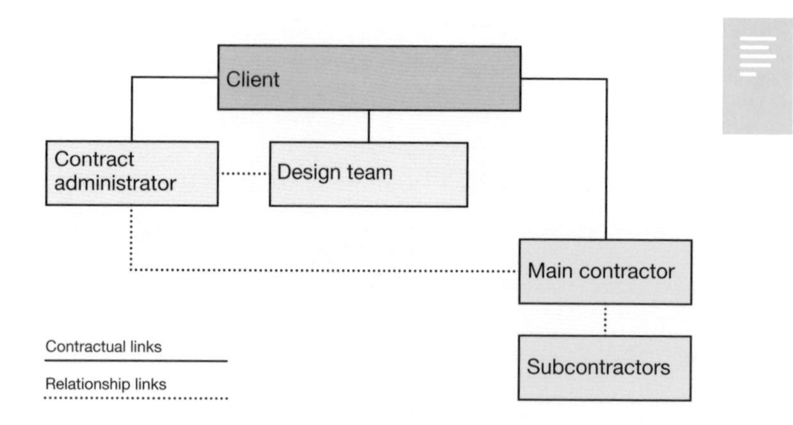

Figure 1.1 Contractual tree for traditional procurement methods

contract administrator is the key point of contact between the client and the contractor.

The client has a direct relationship with the design team (figure 1.1) and the design team retains a high degree of design control and product specification. The contractor is primarily responsible for the quality of work and materials and for achieving completion by the due date. The contractor has no design input unless specifically stated in the contract (eg by the use of the 'contractor's designed portion').

Full working drawings, specifications and documentation are necessary for tendering purposes, and adequate time is needed for the production of this information by the client's design team.

It is possible by use of the contractor's designed portion to make the contractor responsible for discrete aspects of the design. Typically, the contractor will appoint a specialist subcontractor to undertake the actual design, although the main contractor will remain responsible for the design under the terms of the contract. Items that benefit from the design skills of specialist subcontractors include:

- piling
- beam and block floors
- roof trusses
- cladding

- mechanical and electrical services
- other specialist installations.

The extent of the design work relating to the contractor's designed portion is defined by using the Design Responsibility Matrix. With correct use, the Design Responsibility Matrix will show the interface between the level of design input by the Design Team and that undertaken by specialist contractors.

These details should be set out in the preliminaries and the appropriate form of contract used with the contractor's designed portion supplement, eg the JCT Minor Works Building Contract with Contractor's Design 2011 (MWD11).

However, using the 'contractors design portion' form should not be seen as a means of getting the contractor to design major parts of the work; if that is required, the design and build form may be more appropriate even though the same subcontractors may be undertaking the design but in that instance the contractor is responsible for the whole of the design.

This form of procurement requires the client (through the design team) to provide at tender stage a package of information that specifies the works in terms of quality and quantity. This could include drawings, Building Information Modelling (BIM) model, specifications and schedules of work or bills of quantities, although the latter are less frequently used these days. Traditional contracts can be let on the following bases:

- Lump sum: whereby a defined amount of work is undertaken (with or without quantities) for a fixed price that is agreed before construction commences (the contract sum).
- Re-measurement: where the actual quantities are firmed up and costed at the end of the contract by using bills of quantities or some other agreed method – under this arrangement an accurate contract sum will only be known after completion.
- Cost reimbursement: where the contract sum is calculated on the basis of the actual costs of labour, plant and materials to which a fee is added to cover the contractor's overheads and profit. This fee can be a percentage, fixed or fluctuating fee or cost reimbursement based on a target cost. The latter is a variant where the fee is related to an agreed target cost and the fee earned will be affected by the actual cost being higher or lower than the target figure.

Change Control Procedures need to be robust as all variations must be instructed by the contract administrator and should, ideally, be agreed in scope, detail and cost before the work is put in hand. This will allow their impact on the overall contract sum to be monitored as the work progresses.

The contract administrator remains the main point of contact for the client, although in undertaking the duties attached to the role the contract administrator acts independently, as a quasi-arbitrator. The contract administrator is responsible for issuing instructions, inspecting work and ensuring that work is carried out in accordance with the drawings and specifications as well as deciding on claims for delay, authorising interim and final payments and stating when the building is complete.

The traditional route is not the best option for fast-track projects and relies on mutual trust as disputes and problems can have cost and time implications.

What forms of design and build procurement are there?

The design and build form of procurement is used for projects where the client requires time and cost certainty. The client employs a contractor who is responsible for both the design and the construction of the project, acting as a single point of responsibility to the client.

This route enables the client to have certainty on the contract sum, but it is not appropriate for projects where there is a degree of uncertainty; for example, where the brief is unclear or is under development or where the client may wish to make changes during the period of the works. It is not considered ideal for buildings where a high quality is needed or where clients want to use their own subcontractors.

The particular form of design and build contract will depend on the degree of design work the client wishes to undertake before tenders are invited and the extent to which the contractor remains responsible for the works after completion. The client can have some control over the design element but once the contract is let, the client has little direct control over the development of the contractor's detail design.

What is design and build?

The simplest form of design and build (figure 1.2) is where the client commissions a professional (architect, quantity surveyor, project manager,

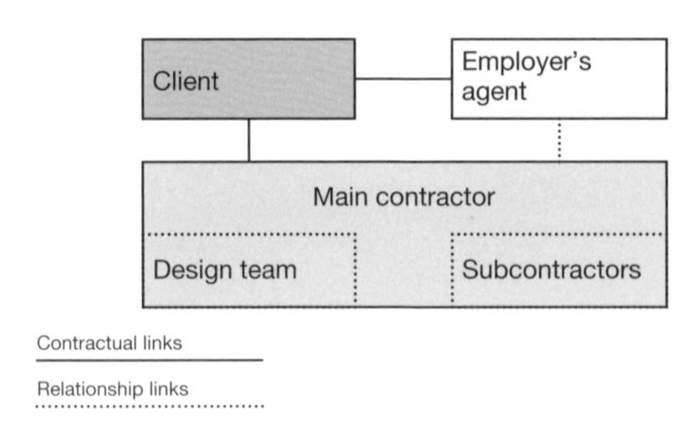

Figure 1.2 Contractual tree for design and build procurement methods

etc.) to act as their agent (the employer's agent). The employer's agent will develop the brief for the building with the client – this will be used to create the Employer's Requirements (see page 61) upon which the prospective contractors will tender. Theoretically, no design work is undertaken by the client and, other than selecting a preferred scheme as part of the tender evaluation, the client has no control over the design. Initial design work will be undertaken by the tenderers in response to the Employer's Requirements – each contractor will employ its own design team to produce an initial proposal, together with an outline specification and a priced schedule; together, these comprise the Contractor's Proposals.

What makes develop and construct different?

A more common form of design and build is 'develop and construct' (figure 1.3), where the client employs the design team and together they prepare a scheme that meets the client's brief, thus the client has an input into the design of the building. This scheme is then submitted for planning approval and tenders are issued on the basis of the planning drawings together with the outline specification and schedules.

In this instance, the client has the option of retaining their design team to monitor the progress of the work or to act as the contract administrator, in which case the contractor will employ their own design team to carry out the detailed design work. Alternatively the client's design team can be novated to the contractor and they then carry out the detailed design work for the contractor.

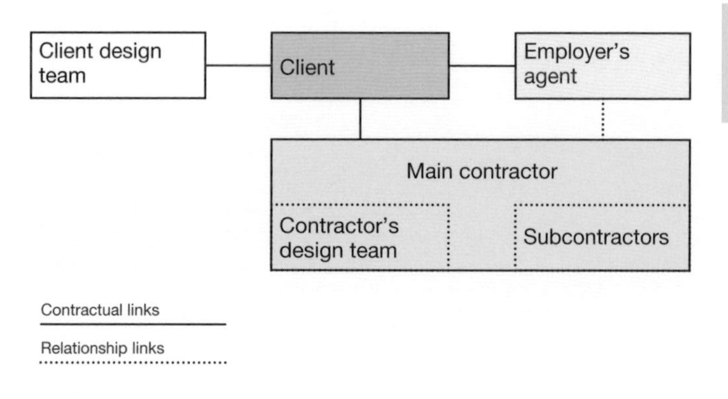

Figure 1.3 Contractual tree for develop and construct procurement methods

What is novation?

Novation is a process by which contractual rights and obligations are transferred from one party to another. On a building project, novation occurs when design consultants who were initially contracted to the client are then 'novated' to the contractor. This is common on design and build projects: the design team are appointed by a client to carry out initial studies or prepare a concept or detailed design, but when a contractor is appointed to carry out or complete the design and construct the works, the design team, or part of it, is novated to work for them.

This is seen to be beneficial to clients as it maintains continuity between pre-tender and post-tender design while leaving sole responsibility for designing and building the project with the contractor.

Novation is not without its problems and consultants often get a feeling of mixed loyalties. A major difficulty is that clients will often feel that the novated design team remain 'their' consultants and that they should continue to do what the client says, whereas in reality the consultants will be working to the instructions of the contractor. In such an instance it is important that the contractual position is fully explained to the client at an early stage, when the procurement route is being discussed. There can also be difficulty

What is novation? (*continued*)

determining where liability rests for design work carried out before novation.

With any novation it is important that the novation documentation is properly drawn up and makes clear which services consultants performed for the client and which services will now be performed for the contractor.

What is design, construct, manage (and maintain)?

A further development of design and build is where a totally integrated service is provided by a single source, which will design, build and maintain the building as a turnkey operation. This can be either contractor or professional led.

What about public sector procurement?

On public sector contracts, when considering the type of procurement and form of contract to use, the contract administrator will need to be guided by European legislation. The UK and the rest of the EU are governed by a number of directives and regulations, which are implemented in national legislation that apply for all public sector projects above a certain value (including consultancy contracts) – see www.ojec.com/Directives.aspx

The directives set out detailed procedures for the award of contracts whose value equals or exceeds specific thresholds, which have to be advertised in the *Official Journal of the European Union* (*OJEU*). Thresholds vary according to the type of work or service – see www.ojeu.eu/Threshholds.aspx. Indeed, the Government Construction Strategy recommends that public projects adopt design and build, private finance initiative or prime-type contract procurement routes, as these are considered to be more collaborative. As a result, it suggests the adoption of the NEC3 form of contract (see page 47), which is believed to encourage collaboration more effectively than some other more traditional contracts, which can be seen as adversarial.

The aim is to deliver high performing buildings. The project team therefore becomes focused on the life-cycle of the building fabric and the systems and there is an increased emphasis on the reduction of long-term operating costs. The perceived benefits to building owners are:

- improved building performance
- sustainability
- reduced operating costs
- reduced financial risk
- reduced responsibility for commissioning of systems
- less operational risk.

What types of management contracting are there?

Management contracting relies on the management expertise of the contractor and is suitable for large projects with complex requirements. Management contracts rely on teamwork and trust and provide the maximum overlap of design, procurement and construction, thus making them very effective. There are many variants, but they generally fall into two main categories: management contracting and construction management. In both instances the contractor is appointed at an early stage, but the appointment should be made at the end of Stage 2 at the very latest for this to work efficiently.

Management contracts are geared towards large-scale, fast-track projects, where an early start and completion are essential factors and the full design information may not be available at the outset. This approach gives the client a degree of flexibility in design, in that the client can make changes during the construction process knowing the full implication of those changes beforehand.

Management contracts are not, however, for inexperienced clients, as the client needs to play an active role in the whole process from beginning to end. Clients will usually have their own in-house experts as well as employing their own design teams.

What is construction management?

In construction management there are no direct contractual links between the construction manager and the trade contractors. Although the trade contractors are organised and administered by the construction manager,

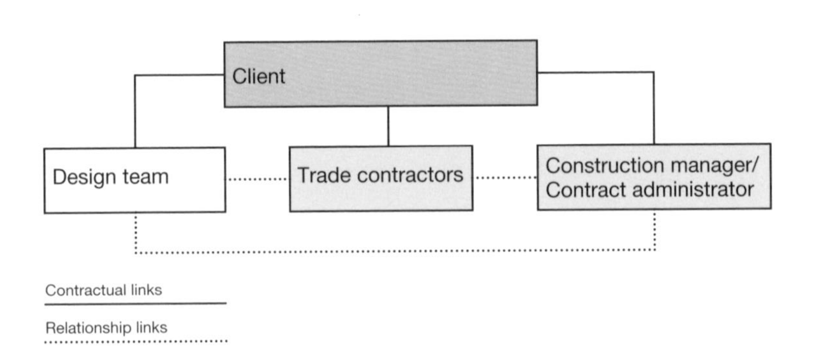

Figure 1.4 Contractual tree for construction management procurement methods

the contractual relationship is between the client and the trade contractors (figure 1.4). The construction manager adopts a consultant's role and is effectively a coordinator working for the client to administrate the contract and act on the client's behalf.

What is management contracting?

In management contracting the client employs the design team. The management contractor is appointed at the initial stages of a project under the direction of the contract administrator (figure 1.5). The management contractor's appointment follows a tender process, which would comprise the submission of a written proposal and details of the management fee, followed by an interview by the client and the design team. The management contractor then effectively becomes a member of the design team, to advise on the design and construction programmes and instigate tender action for the delivery of goods and materials.

The contract is divided into works packages, such as groundworks, brickwork and roofing. Following the tendering of the work packages, the management contractor directly employs and manages the individual trade contractors, who carry out the works and undertake their own design work as appropriate.

During the construction period the management contractor will set out and manage, organise and secure the carrying out and completion of the project through the works contractors who are directly contracted to the management contractor. Responsibility for managing the terms of the

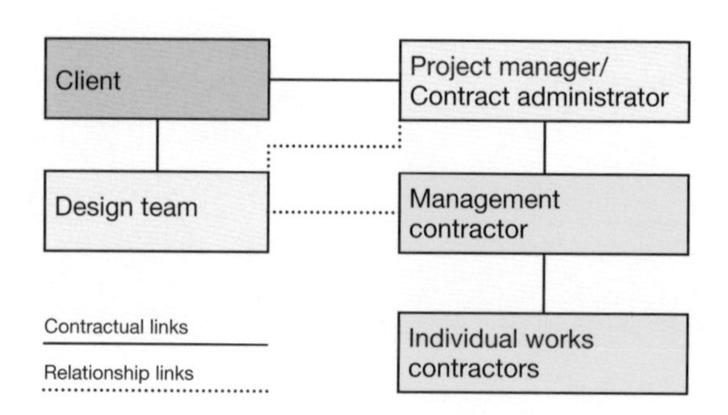

Figure 1.5 Contractual tree for management contracting procurement methods

contract also rests with the management contractor but the consequences of default by a works contractor does not fall on the management contractor provided the contract terms have been complied with.

This form of procurement allows any complex detailing to be undertaken by the specialist trades at an early stage but, given that each individual work packages is tendered, this gives the client the maximum price competition for each package.

What is partnering and collaborative working?

Collaborative procurement, or partnering, is intended to have the mutual objectives of delivering the project on time, to budget and to quality. Partnering was one of the recommendations of the Latham Report in 1994, intended as a way for clients and contractors to work together with the aim of reducing costs and minimising conflict. There needs to be a spirit of trust and mutual cooperation within a team approach – it is about the management of the project rather than a procurement method. It is flexible and can apply to projects of all types and sizes.

The contractual structure is set up such that the contractor and subcontractors, as well as the design team, are appointed as early as possible in the design development process, working in accordance with a single programme to achieve an early commencement on site.

This relationship continues through to the construction stage, so any changes or delay and disruption or any problems in performance will be identified early. That said, should there be any issues, there is a structured approach to alternative dispute resolution, including problem solving and referral to conciliation or mediation.

Partnering contracts, such as PPC2000, integrate the whole project team and avoid duplications between team members, which avoids wasting time and money on resolving issues at a later stage. There are clear timetables through to start on site, with a resultant cost saving, and earlier contractor and specialist input, leading to innovations and efficiencies with the potential to improve quality and reduce cost.

Above all, there is improved performance within the design team, the contractor and specialist subcontractors and suppliers through the early creation of the partnering team, improved communication and mutually compatible roles and responsibilities, with more open Cost Information to establish price accuracy, removal or reduction of price contingencies, and closer control over the consequences of changes and unforeseen events.

Many public sector clients use project partnering, where all parties sign a single multiparty contract. This is also being used on UK government pilot projects following the publication of the Government Construction Strategy in May 2011, which sets out the principles of an alternative approach for procurement, designed to eliminate the wastefulness of several teams completing and costing a series of alternative designs for a single project, only one of which will be used.

How is the procurement route chosen?

It was stated above that there are essentially three basic elements that determine the selection of the procurement method: time, quality and cost. However, these core elements can be expanded to further refine the selection (figure 1.6).

Although the procurement method and form of building contract may have been decided before the contract administrator is appointed, the decision on which procurement method and contract form are to be used will be made at Stage 1. This will depend on:

- the nature and scope of the work
- how risks are apportioned

	POSITIVES	NEGATIVES
TRADITIONAL	Client maintaining control of quality and specification Cost certainty when accepting tender	Client risk Cost control during construction works
DESIGN AND BUILD	Transferring risk to contractor	Loss of control by client in influencing quality and specification
MANAGEMENT	Client retains full control of design Shorter programme possible	No transfer of risk to contractor Risk to client if design is not fully coordinated No cost certainty until later in programme
CONSTRUCTION MANAGEMENT	Client retains full control of design Shorter programme possible	Client employs the package contractors directly (the construction manager has no contractual role), therefore the client carries majority of the risk No transfer of risk to contractor Risk to client if design is not fully coordinated No cost certainty until later in programme

Figure 1.6 Comparison of procurement methods

- how and where the design responsibility is placed between the architect, consultants, contractor and subcontractors
- the basis upon which the price is to be set (whether fixed or variable)
- the balance between time, cost and quality.

How is risk considered?

Some degree of risk has to be taken in all forms of building contract. There are, of course, insurable risks – injury, fire and so on – but it is the 'speculative risk', which we are primarily considering when looking at the procurement route.

This speculative risk needs to be apportioned between the parties from the outset and should relate to unforeseen matters, such as unexpected ground conditions, bad weather and other matters beyond anyone's control, which could result in losses for one of the parties in terms of

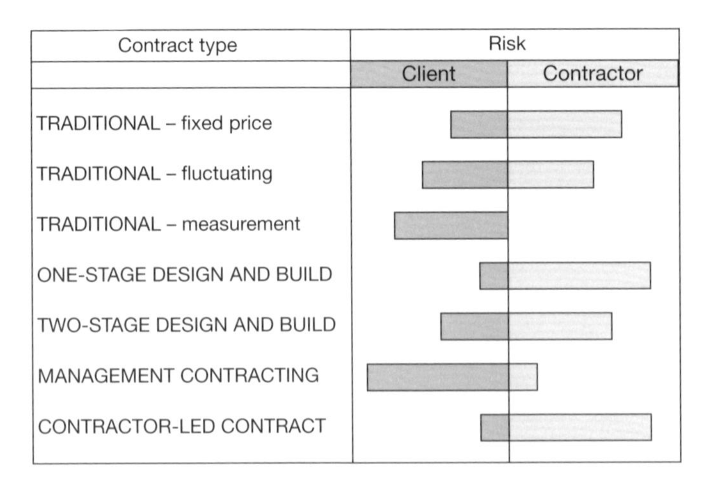

Figure 1.7 Speculative risk table

time and money. The level of this risk must be assessed at the outset and, more importantly, it must be clear how this risk is apportioned between the parties.

As can be seen from figure 1.7, the balance of risk between the client and the contractor varies depending on which procurement route is under consideration. The traditional route gives a relatively equal balance of risk between the client and the contractor, whereas the risk in a design and build contract lies almost wholly with the contractor. In contrast, the risk is almost wholly with the client in management contracts, where the extent of the work and project cost are not known at the outset.

When considering the procurement route, thought should also be given to the form of Building Contract to be used.

What forms of building contract are suitable?

It is essential that on any project, regardless of size, a contract exists between the client and the contractor. The Building Contract sets out how the parties will deal with issues such as payments, variations, insurances, etc. Different forms of contract impose different conditions on the parties, therefore it is important that the correct form of Building Contract is used with the appropriate procurement route. The form of Building Contract should preferably be aligned with the professional services contracts used for the design team appointments, consideration should be given to using professional services contracts issued by the same body as the main building contract.

Where do I find out more about Building Contracts?

It is beyond the scope of this guide to discuss Building Contracts in detail. For a comprehensive guide to contacts, refer to *Which Contract?* by Hugh Clamp, Stanley Cox, Sarah Lupton and Koko Udom (RIBA Publishing).

What is a contract?

A contract is a legally binding agreement which sets out the responsibilities of the parties, commits the parties to performing the terms of the contract and allocates risks between them. It is normally in two parts, comprising the articles of agreement and the conditions of contract.

A building contract is a contract between a client and a contractor for construction works. On some projects there may be more than one contract; for example, on a commercial building there may be one contract for the shell and core and a separate one for the fit-out.

A contract is made by agreement, consideration and intention, and can even be made verbally, by fax, email or letter.

While a verbal agreement effectively becomes a contract, it can be difficult to prove, therefore a tender and a letter of acceptance are more appropriate for building works. Indeed, the better the form, the fewer difficulties may arise.

The progress of a contract will go through four distinct stages:

- pre-contract
- preparing of the contract details
- performing the works
- completing the works.

Contract completion occurs when the parties have performed or discharged all their obligations completely.

It is often believed that an invitation to tender is effectively an offer of a contract, but it should be noted that in common law an invitation to tender is *not an offer* but an invitation to negotiate.

What forms of Building Contract are there?

Building Contracts have been developed to suit each procurement route. The preference should be to use standard forms (rather than bespoke contracts) as they are more convenient, less expensive and guidance is available.

The Latham Report recommended using standard forms without amendment. Even the slightest amendment could affect the balance and precise meaning of the agreement, which potentially has serious implications if tested in law. Therefore, ad hoc amendments should be avoided, particularly on points of substance. Should an amendment be considered necessary, it should be done only with appropriate professional advice.

Where non-standard agreements are used they are prepared by lawyers. Such agreements are primarily used on commercial and large, complex projects.

Over 40 standard forms of Building Contract are available. The vast majority have been produced by the Joint Contracts Tribunal (JCT), but several other organisations and forums have prepared alternative forms. The different standard forms are considered below.

What standard forms of Building Contract are there?

Many options for building contract forms are currently published by various bodies for use with alternative procurement routes: traditional, design and build and so on. There are simpler versions for smaller projects, and variants that allow for contractors to design specific elements of the works. It should be noted, however, that this guide has not examined every form of contract or variant in detail, but has highlighted the range and principles of the alternative procurement routes.

Links to building contract bodies

RIBA	Royal Institute of British Architects (www.ribacontracts.com)
JCT	Joint Contracts Tribunal Ltd (www.jctltd.co.uk)
GC	The UK Government Stationery Office (www.hmso.gov.uk)
NEC	The Institution of Civil Engineers: New Engineering Contract (www.neccontract.com)
ACA	The Association of Consultant Architects (www.acarchitects.co.uk)
CIOB	Chartered Institute of Building (www.ciob.org/contract-complex-projects)
FMB	Federation of Master Builders (www.fmb.org.uk)
SBCC	Scottish Building Contract Committee (www.sbconline.com)
ICC	The Association for Consultancy and Engineering and the Civil Engineering Contractors Association (www.rics.org)
FIDIC	Fédération Internationale des Ingénieurs-Conseil (International Federation of Consulting Engineers) (www.fidic.org)

RIBA contracts

The RIBA has now developed two new Building Contracts to cover matters not properly covered by existing contracts, known as the Domestic and Concise Building Contracts.

Historically, it has been common practice on domestic projects to use a commercial contract, but because of the nature of these some of the contractual terms need to be individually negotiated with the consumer client; failure to keep to this requirement could lead to misunderstanding and expensive disputes.

Therefore, the RIBA Domestic Building Contract provides a simple yet comprehensive contract solution for building works at a client's home. It is suitable for all types of domestic works, including simple renovations as well as more complex extensions and new buildings.

The RIBA Concise Building Contract is the commercial version, suitable for all types of small and minor commercial building work. It is written in plain English for easy understanding and is shorter than other commercial forms.

In both instances there are optional clauses to cover more advanced contractual terms while retaining their overall simplicity.

JCT contracts

The Joint Contracts Tribunal (JCT) has produced standard forms of contract, guidance notes and other standard documentation for use in the construction industry since it was established in 1931 by the RIBA and the National Federation of Building Trades Employers (NFBTE). Since then it has expanded the number of contributing organisations and it provides a wide range of forms of contract to meet the various and diverse needs of UK construction. JCT contracts are intended to be fair and evenly balanced between the parties.

NEC forms of contract

The NEC is a family of standard contracts that was originally launched in 1993 as the New Engineering Contract, which looked towards a collaborative and integrated working approach. NEC3 is the third edition, launched in 2005. NEC contracts are suitable for procuring a wide range of UK and international building works and services, from major framework projects through to minor works and the purchasing of supplies and goods. As well as building contracts, the NEC3 suite includes compatible professional services contracts.

The objectives behind the NEC were to allow flexibility in use, to be clear through the use of ordinary language (minimising legal language), to be a stimulus to good management and to encourage collaborative working relations between the two contracting parties. The implementation of NEC contracts has resulted in major benefits for projects both nationally and internationally in terms of time and cost savings and improved quality.

NEC3 contracts include a range of optional clauses that can be used to tailor the contract terms to the particular type of work or services. The principal contracts comprise the core clauses, which contain the common terms, together with a choice of six options. Users can choose the option deemed to be the most suitable to give the client the best option/value for money on the project.

CIOB contracts

Chartered Institute of Building contracts are intended for large, complex projects with experienced clients. The contracts, which adopt a plain English approach, are used internationally, particularly where FIDIC contracts are considered overcomplicated and where JCT is not selected.

FIDIC forms of contract

These contracts are commonly used where tenders are invited on an international basis, particularly in countries where standard forms do not exist or are unsuitable. The English versions of all the contract forms are considered to be the official and authentic version.

The International Federation of Consulting Engineers (FIDIC, an acronym for its French name, *Fédération Internationale des Ingénieurs-Conseil*) is an international standards organisation for the construction industry which promotes and implements international standard forms of contracts for works and for clients, consultants, sub-consultants, joint ventures and representatives.

In 1999 FIDIC published a new suite of standard forms of contract which are suitable for the great majority of international construction and plant installation projects around the world. The various forms are known by their book colour, Red, Yellow, Green or Silver.

Project partnering contract PPC2000

PPC2000, published by the Association of Consultant Architects, is a form of multiparty contract for the procurement of capital projects. The key differences between PPC2000 and any other published contract form are that it:

- integrates all members of the partnering team under a single multiparty contract

- covers the entire duration of the design, supply and construction process
- includes team-based timetables, controls and problem-solving mechanisms.

In this form of contract the consultant team, the contractor and specialist subcontractors and suppliers all operate under the same terms and conditions so that they are fully aware of each other's roles and responsibilities, with each owing a direct duty of care to the others.

Selecting the form of contract

When making a decision on which form of Building Contract is to be used, the contract administrator has to consider two things: first, the procurement route, and, second, the type of contract, eg JCT, NEC.

It may be that the client has a preference for a particular form of contract, based on experience from previous projects, or that government rules dictate the use of a particular form. However, many clients will need the contract administrator to guide them as part of the Stage 1 outputs.

As the contract administrator may not have been appointed at this stage, the project lead or other project team member contributing to the briefing process should give full consideration to the future role of the contract administrator and ensure that questions are asked or answered on the contract administrator's behalf.

Guidance on contract selection

The choice of Building Contract form will be guided by the procurement route and further guidance can be obtained directly from the relevant contracts bodies. For example, the JCT produces a *Practice Note – Deciding on the Appropriate JCT Contract 2011,* which includes a matrix to assist in selecting the appropriate form of contract to suit a particular procurement route or variation.

What are collateral warranties and third party rights?

When advising the client on contract forms the contract administrator should consider the need for any further warranties between the parties, whether contractor or professional, to ensure that the client or any future purchasers are protected. As well as their use between contractors and subcontractors, these can be between clients, consultants, funders, purchasers and tenants to warrant that they have fulfilled their duties under the building contract.

The contract particulars should set out the details, but the obligations should be no more onerous than those in the primary contract. Standard forms of collateral warranty are available, such as those prepared by the British Property Federation, the Construction Industry Council and the JCT.

An alternative to collateral warranties available through section 1(1) of the Third Party Rights Act 1999 is the use of 'third party rights', which is the right of a person who is not a party to a contract (the third party) to enforce the benefit of a term of that contract, such as a funder enforcing a term of a contract between the client and a professional or the contractor.

As with collateral warranties, third party rights provide construction security if something goes wrong on a project. For example, a tenant might want to be able to claim for loss directly from the person who caused the loss, such as the architect or the contractor.

Without either a collateral warranty or third party rights, a tenant may be unable to make an effective claim because a claim for loss outside the law of contract (tort) is unlikely to succeed.

In practice there is no difference between collateral warranties and third party rights as both can be effective. However, collateral warranties are more common as they are more familiar to the parties and, once signed, are contracts like any other.

What procurement activity is there during Stage 1?

Contractor-led involvement at the start of Stage 2: Concept Design

In the case of a contractor-led contract – where the contractor is to lead the design process – it is essential that the contractor is ready to commence

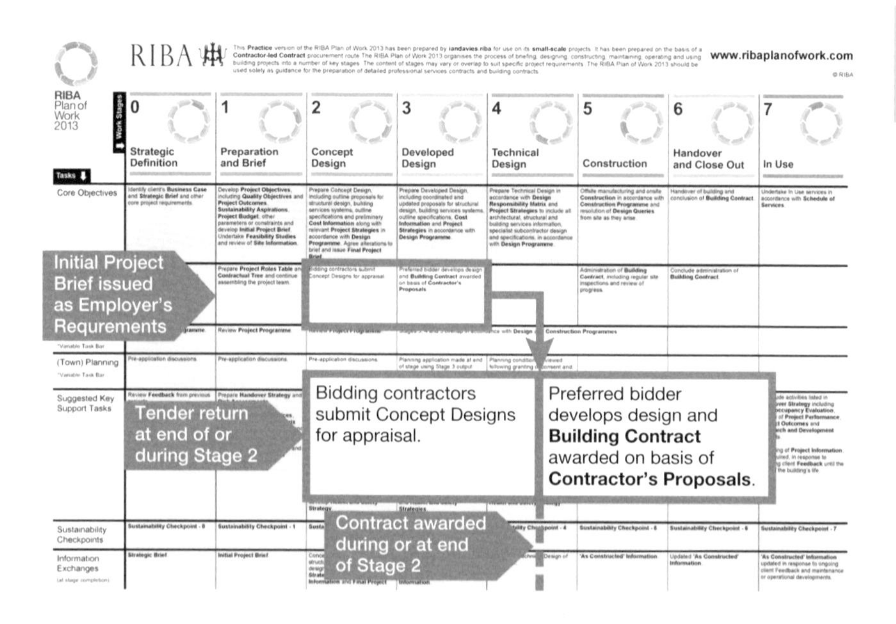

Figure 1.8 Bespoke Plan of Work for a contractor-led project

work on the design and tender in earnest at the beginning of Stage 2. Figure 1.8 shows a bespoke Plan of Work for a contractor-led project where the design and tendering is undertaken at Stage 2. Therefore, in order to achieve this, the contract administrator should issue the tender documents at the end of Stage 1.

Tender documents at Stage 1

At the end of Stage 1 the Initial Project Brief will have been prepared and this will be issued as part of the Employer's Requirements within the tender

Guidance on tendering

Further information on tendering can be obtained from the JCT *Practice Note – Tendering 2012* or from *The Guide to Tendering: Construction Projects*, by the NBS. www.thenbs.com

documentation. This will be issued to a number of shortlisted contractors to enable them to prepare their concept design and initial tender.

The principles of tendering and tender documentation are dealt with in more detail at Stage 4.

Chapter summary 1

This chapter has demonstrated the principles to be considered when selecting the appropriate procurement route and form of building contract for any building project.

The project strategy will necessitate making an analysis of the situation, making a choice from the procurement options and then devising a method of implementing that choice, using well-established rules and procedures. The client's policies, available resources, organisational structure and preferred contractual arrangements will need to be taken into account. There are inherent risks associated with using any particular procurement strategy, but equally important is the need for all parties to comply with their respective obligations; this is particularly important where responsibilities for design and construction are separated.

Although the contract administrator may not have been appointed at this stage, the project lead or other team member contributing to the briefing process will need to give full consideration to the future role of the contract administrator and questions to be asked or answered on their behalf.

In order for the Concept Design work to be started in earnest in Stage 2, the tender documentation for contractor-led projects should be issued at the end of Stage 1.

Stage 2

Concept Design

Chapter overview

Stage 2 is crucial on any project as the Concept Design will be developed and signed off by the client. At Stage 2 the contract administrator would not necessarily have been appointed for some projects as the timing of this appointment is dependent on the procurement route selected at Stage 1 which, along with the Construction Strategy, has to be reviewed at this stage.

Tendering occurs on some procurement routes and this chapter considers how the contract administrator takes the information prepared by the design team as appropriate to the stage and assembles it with other documents prepared by the contract administrator to enable competitive prices to be obtained for the work from selected contractors.

However, the forms and procedures for inviting and receiving tenders together with tasks associated with the contract documents immediately prior to the awarding of the contract are dealt with in more detail in Stage 4.

The key coverage in this chapter is as follows:

What procurement activity is there at Stage 2?

What tendering activity is there at Stage 2?

Introduction

The priority for the contract administrator during this stage will be to facilitate the successful receipt of competitive prices for the works appropriate to the procurement route selected, so that the client can decide who should develop the detail and, ultimately, build their project. The contract administrator should advise the client on the preparation of a satisfactory list of suitable contractors and take the client through the tender process before advising on the most suitable contractor.

The level of detail which accompanies the tender documents will vary depending on the level of design control that the client wishes to retain and the level of detail to be developed by the main contractor or specialist subcontractors at Stage 4. Early in the stage the contract administrator should discuss the job-specific tender procedures with the client so that the client is fully aware of the process and the implications for the contractors and the project.

Tendering for some procurement routes will occur at Stage 2. As the principles and procedures of tendering are common to all procurement routes, the role of the contract administrator in respect of tasks and procedures in the selective tendering process will be described in more detail at Stage 4.

What are the Core Objectives of this stage?

The Core Objectives of the RIBA Plan of Work 2013 at Stage 2 are:

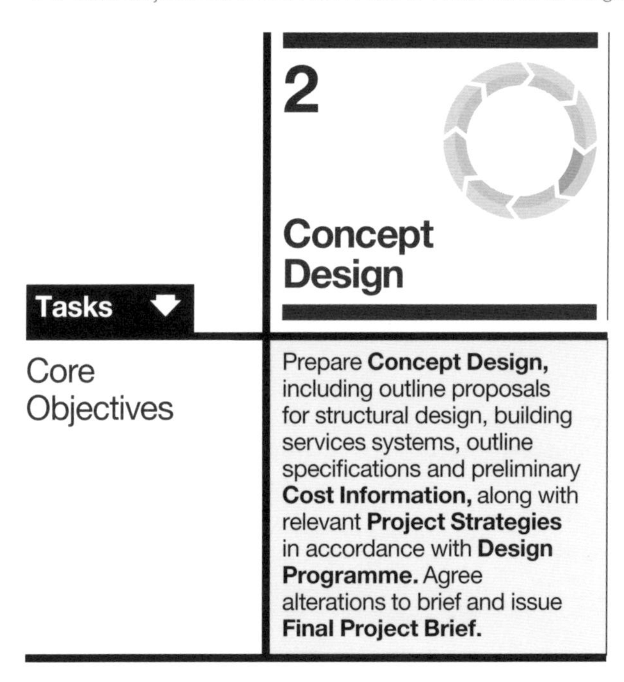

The Core Objectives at Stage 2 comprise the preparation of the Concept Design and preliminary Cost Information. The Initial Project Brief will be developed and alterations agreed to create the Final Project Brief. The design team will be required to complete its design in sufficient detail to suit the procurement route and to enable competitive tenders to be obtained, leading to the award of the contract to the preferred contractor.

What supporting tasks should be undertaken during Stage 2?

The Suggested Key Support Tasks noted in the RIBA Plan of Work 2013 have been devised to support the Core Objectives and to ensure that the documentation required to proceed to the next stage has been prepared. At this stage the suggested tasks applicable to the role of the contract administrator are:

- Prepare Maintenance and Operational Strategy and review Handover Strategy and Risk Assessments.
- Review and update Project Execution Plan.
- Consider Construction Strategy, including offsite fabrication, and develop Health and Safety Strategy.

The support tasks during this stage are focused on strategic matters and the preparation of various key strategies. However, within a bespoke Plan of Work the variable procurement task bar will fix the tendering-related support tasks for the contract administrator at Stage 2 based on the procurement route selected.

What procurement activity is there at Stage 2?

Two-stage design and build projects will be tendered at this stage based on the level of information available at the time to select a preferred contractor for the pre-construction services. Management contracts will also be tendered during Stage 2, ready for the contract to be awarded at the end of the stage.

In the case of a contractor-led project the tender documentation will have been issued to a number of shortlisted contractors at the end of Stage 1 to enable each to prepare a concept design and initial tender. During Stage 2, the contractor's design is developed, leading to their concept scheme and tender being submitted for appraisal by the client and the design team based on the selected criteria at the end of the stage.

What tendering activity is there at Stage 2?

This section details the variations in the tender documentation for Stage 2 tender activities for two-stage design and build and management contracts.

Two-stage design and build

Tenders are most commonly invited on a single-stage basis, but a two-stage process may be more appropriate on larger more complex projects where it is seen that there is an advantage in the contractor collaborating in the development of the design.

At Stage 2 (Concept Design) the project will be tendered on the basis of the level of information available to select the preferred contractor for the pre-construction services. During Stage 3 (Developed Design) and Stage 4 (Technical Design) the contractor will work with the design team in the development of the working drawings, specifications and schedules to produce the Contractor's Proposals. These will then be priced incrementally and, if the price is satisfactory, the contract for construction is awarded during Stage 4.

Preliminary enquiry letters should be sent to each prospective tenderer giving the outline of the project and the tendering details. Once the

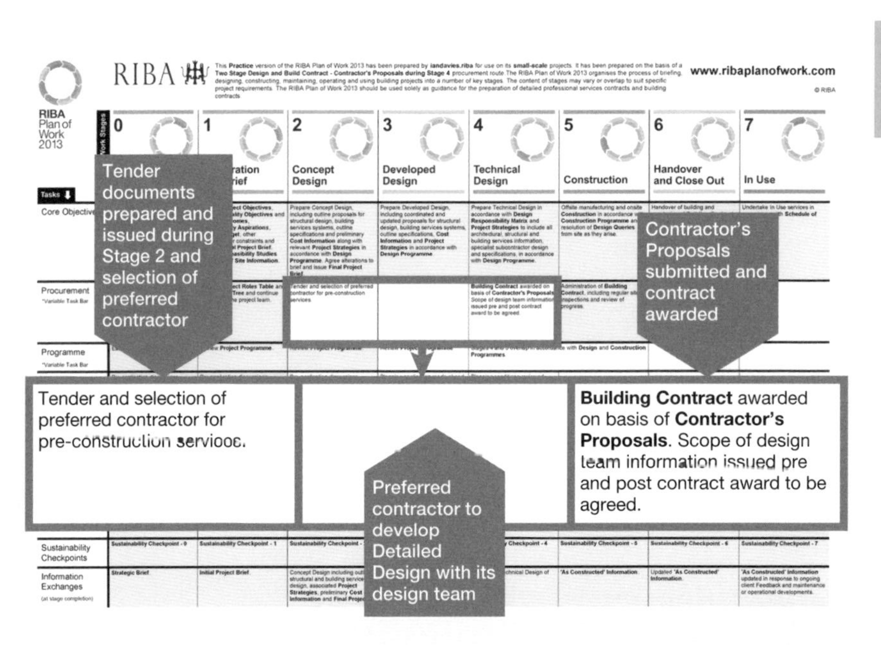

Figure 2.1 Bespoke RIBA Plan of Work 2013 for a two-stage design and build project

tenderers have confirmed that they are interested, a final selection can be made and references taken up as appropriate.

On design and build contracts the tender documents are usually referred to as the Employer's Requirements and differ slightly in the level of detail and content from a traditional contract.

What are the Employer's Requirements?

The Employer's Requirements is the term commonly applied to the main tender document used on design and build projects. This document conveys to tenderers the details of the client's requirements, the specification for the building and the scope of services required from the contractor. The level of detail to be included can vary significantly and will depend on whether the requirements are for a one- or two-stage process.

The level of detail could range from a site plan and schedule of accommodation on a simple design and build project, to a full design up to planning stage on a develop and construct project. An indicative list of items which may be included is shown at the appropriate stage for the selected procurement route.

The detail within the document will depend on how much design control the client wants to retain and how far the contractor will develop the design detail.

The Employer's Requirements for a two-stage tender document should include the following items (the list is not exhaustive):

- a project overview
- the scope of services required, including identification of elements requiring contractor design
- the form of the Contractor's Proposals required
- the format required for the contract sum analysis
- site details
- site constraints
- outline planning permission and conditions, if known
- functionality of the building
- schematic layouts/flow charts
- schedules of accommodation

- Project Programme information
- details of the contract, preliminaries, insurances, etc.
- schedules of rates and fees to be included in the tender proposal
- tender return form.

The contract administrator will assemble and issue the various tender documents and arrange for the receipt of the tenders, either by hand or electronically. The tender process must be fair, open and confidential, and the contract administrator must ensure that the client is fully aware of this.

On traditional projects, the contractor is usually selected on the basis of the lowest priced tender. On design and build and management contracts, however, the client will often decide on which contractor to use by carrying out an interview with the lowest or highest scoring contractor and their team. The purpose of this is twofold: first, to interrogate the contractor's design where one has been produced and, second, to ensure that the contractor's team will be compatible with the rest of the project team and to understand the way in which the contractor intends to manage the contract. Once the contractor is decided, the contract administrator should advise the successful bidder of the intention to place the contract. The contract documents will then be prepared by the contract administrator for signature before the works commence.

The tender documents prepared at Stage 2 invite tenders for the first stage to be submitted on the basis of a fee for prescribed pre-construction services, preliminaries profit and overheads. The level of detail issued with the documents will depend on the level of design control the client wishes to retain. This can range from the contractor being given a completely free hand based on an accommodation schedule, through to the issue of Feasibility Studies for the contractor to develop.

Following the preparation of the tender documents the contract administrator should follow the general rules and principles for the selection and tendering process, as detailed in Stage 4.

At the end of Stage 2 the preferred contractor will be commissioned to undertake the pre-construction works only. The contractor will be given a limited appointment to start work on collaborating with the project team in the development of the design. The contract administrator should confirm this in writing pending the issue of a formal letter of intent at the end of Stage 4, immediately prior to the award of the Building Contract after the design has been developed in detail and the contractor has submitted priced Contractor's Proposals.

The main benefit of the contractor being involved before the design is finished is that the contractor can contribute to the final design and the second stage tendering process. However, there can be no guarantee for the contractor that the client will proceed to construction after the design stage.

Management contracting

The Building Contract for a management contract will be tendered during Stage 2 and the contract will be awarded at the end of the stage. Subsequent design information based on agreed work packages will be developed and tendered during Stages 4 and 5, thus achieving the maximum overlap of design and construction.

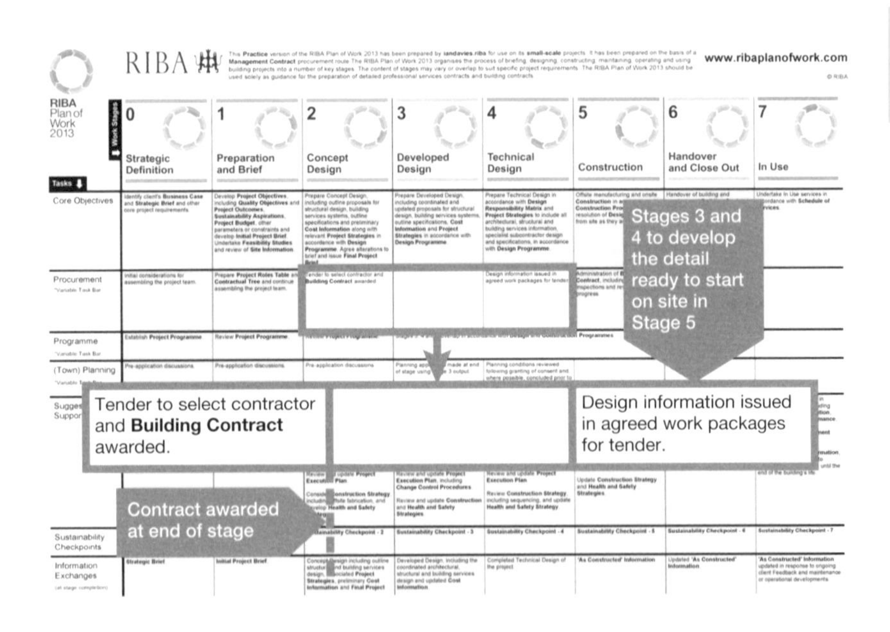

Figure 2.2 Bespoke RIBA Plan of Work 2013 for a management project

As previously outlined (at Stage 0), management contracts generally fall into two main categories, management contracting and construction management. In both cases the contract administrator should pre-agree a shortlist of contractors or construction managers with the client and the design team. In the case of management contracts, the client is usually experienced in construction practice and procedures and is likely to have their own approved list.

Construction management

The construction manager is employed in the same way as any of the other consultants within the design team. Unlike the contractor, who is selected through a tendering process, the construction manager will be selected by the client on the basis of experience, knowledge and fees.

Management contracting

On management contracts, given the early stage in the construction process at which the Building Contract is awarded, the level of detail issued with the tender documents will depend on what information is available at the time.

The Employer's Requirements, which should accompany the tender documents, should include the following items (the list is not exhaustive):

- site details
- site constraints
- any topographical surveys
- any geotechnical reports
- outline planning permission and conditions, if known
- any statutory consultations held
- existing health and safety files, if appropriate
- functionality of the building
- schematic layouts/flow charts
- schedules of accommodation
- room data sheets
- specification information
- equipment requirements
- programme information.

Following the preparation of the tender documents, the contract administrator should follow the general rules and principles for the selection and tendering process as detailed for Stage 4.

The prospective management contractors should be asked to submit a written proposal outlining:

- their approach to the project
- health and safety information
- a Schedule of Services
- prime cost allowances for the supply of necessary items not yet finally selected, eg doors and windows
- their management fee.

The contract administrator should evaluate the tenders, possibly using a cost/quality matrix determined by the client or project manager, and prepare a shortlist of contractors to invite for interview. At the interviews each contractor should be given the opportunity to expand on their proposal, ask questions, answer queries and introduce their project team. The importance of the latter should not be underestimated as it is key that the project team should be able to work together in a collaborative way.

At this point the Building Contract is awarded to the successful management contractor following which the contract administrator should carry out the procedures and preparation of the contract documents detailed in Stage 4.

Chapter summary 2

This chapter has described tender tasks and procedures required at Stage 2 for two-stage design and build and management contracts, where the contractor is appointed early in order to work with the design team to develop the Concept Design.

At this stage, the design for any procurement route will be in its infancy, and other members of the project team will be undertaking their respective roles during this period. However, it is the contract administrator who has the key role in ensuring that fair and competitive tenders are obtained so that the client can identify the most suitable contractor with whom to develop the scheme. Throughout this process the contract administrator needs to be methodical in the procedures to be undertaken, accurate and thorough.

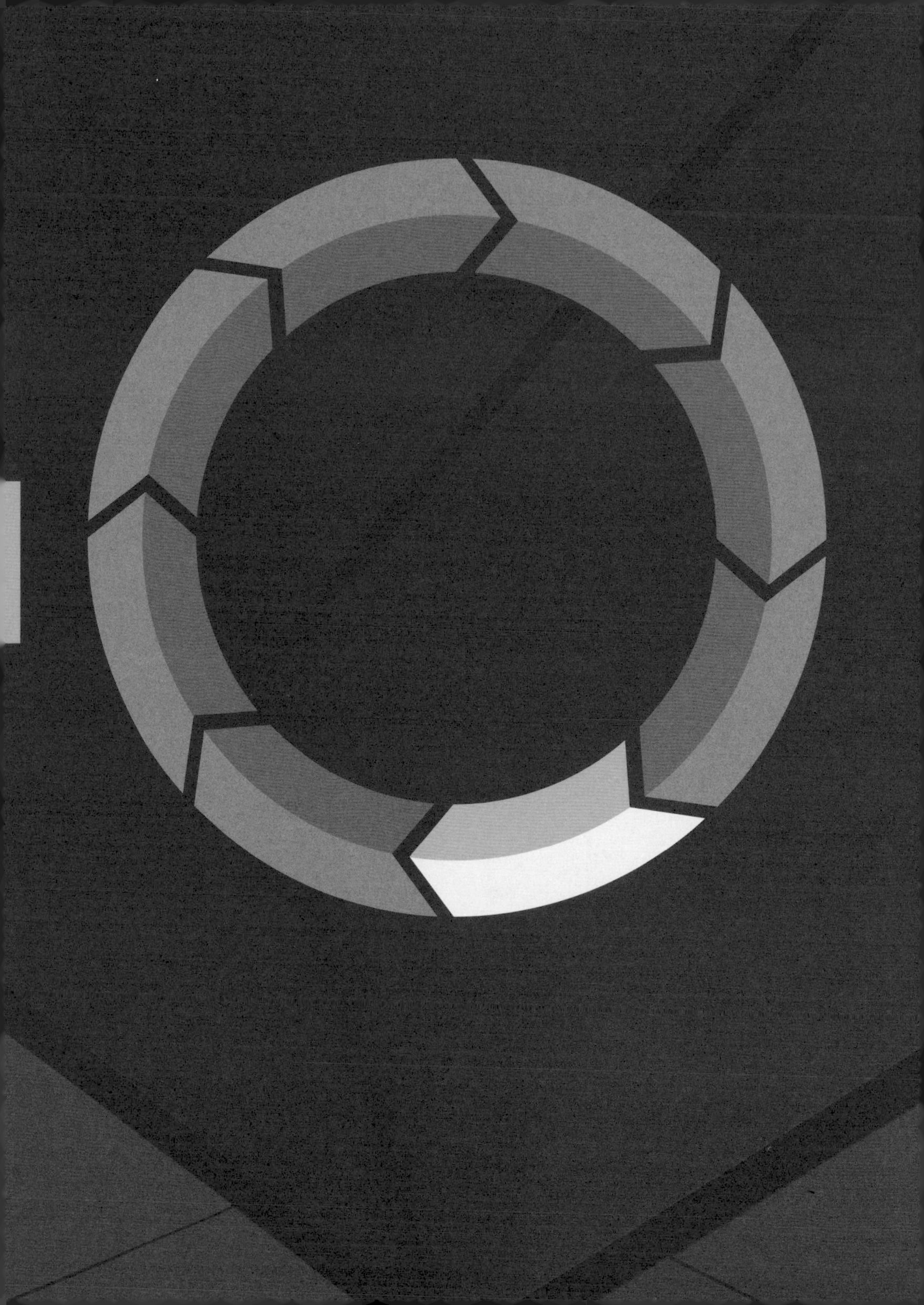

Stage 3

Developed Design

Chapter overview

At Stage 3 the role of the contract administrator continues to be dependent on the procurement route selected at Stage 1.

During Stage 3 the Concept Design will be further developed through input from other members of the design team, to the point where a planning application could be made. There may be various iterations of the design during this period, but by the end of the stage the lead designer should have coordinated the scheme and arranged for it to be signed off by the client.

This chapter looks at what tasks the contract administrator would undertake at this stage and the differences between the tender documents required for one-stage design and build projects.

The key coverage in this chapter is as follows:

What procurement activity is there at Stage 3?

What tendering activity is there at Stage 3?

Introduction

During this stage the contract administrator will facilitate the successful receipt of competitive tenders for a one-stage design and build project so that the client can decide who should develop the detail and ultimately build the project.

Early in the stage the contract administrator should discuss the job-specific tender procedures with the client, so that the client is fully aware of the process and the implications for the contractors and the project.

The level of detail in the tender documents issued at this stage will be higher than in tender documents issued at an earlier stage as the design will be more developed. The role of the contract administrator in the selective tendering process together with works associated with the contract documents immediately prior to the awarding of the contract, will be similar to the traditional procurement route. These are described in the following chapter (Stage 4).

What are the Core Objectives of this stage?

The Core Objectives of the RIBA Plan of Work 2013 at Stage 3 are:

The Core Objective at Stage 3 is to develop the Concept Design to a stage where the design is coordinated and is suitable for making a planning application as appropriate. The design team will be required to complete the Developed Design such that it reflects the increased level of detail required and, depending on the procurement route, enables competitive tenders to be sought and the Building Contract awarded to the preferred contractor.

What supporting tasks should be undertaken during Stage 3?

The Suggested Key Support Tasks noted in the RIBA Plan of Work 2013 have been devised to support the Core Objectives and to ensure that the documentation required to proceed to the next stage has been prepared. At this stage, the suggested tasks applicable to the role of the contract administrator are:

- Review and update Maintenance and Operational and Handover Strategies.
- Review and update Project Execution Plan, including Change Control Procedures.
- Review and update Construction and Health and Safety Strategies.

However, within a bespoke Plan of Work the variable procurement task bar will fix the tendering-related support tasks for the contract administrator at Stage 3 based on the procurement route selected. This will generate key outputs, which will include the tender documents to be used in the tender process, leading to the award of a contract to the successful contractor.

What procurement activity is there at Stage 3?

One-stage design and build projects will be tendered at this stage.

At Stage 3 (Developed Design) there are no tender actions for traditional and two-stage design and build, but in the case of the latter, the selected tenderer (recommended as the 'preferred contractor') would be working to develop the design ready for submission as part of the Contractor's Proposals at the end of Stage 4.

On management contracts the Building Contract will have been awarded at the end of Stage 2. During Stage 3, the detailed design information will be developed into separate works packages, to be tendered as part of their Stage 4 operations to meet the contractor's Construction Programme.

On a contractor-led project the preferred contractor will be developing their design and proposals, with the Building Contract being awarded either during or at the end of Stage 3.

What tendering activity is there at Stage 3?

Procurement activity at Stage 3 will be restricted to tendering activity for one-stage design and build and the contract award for contractor-led procurement routes. This section summarises the basic principles for tendering at this stage, which are expanded upon in more detail in the following chapter (Stage 4).

As well as ensuring that the tender process is fair and clear for all parties, the contract administrator needs to be confident that each contractor chosen to tender will be able to undertake the work. To achieve this, the contract administrator should first prepare a list of pre-qualified contractors (see Stage 4 page 91)

One-stage design and build

At Stage 3 the contract administrator will use the information produced for the project to prepare the Employer's Requirements and then issue

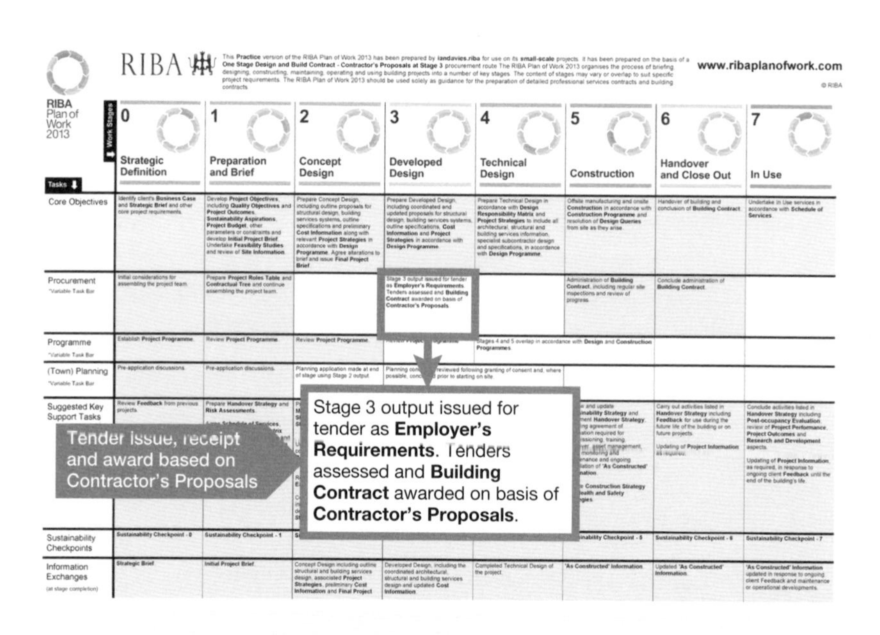

Figure 3.1 Bespoke RIBA Plan of Work 2013 for a one-stage design and build project

the tender documents. Following receipt and assessment of tenders, the Building Contract will be awarded to the selected contractor at the end of the stage, on the basis of the Contractor's Proposals.

As for other forms of Building Contract, the contract administrator should discuss with the client the method of tendering to be used and its implications to obtain the client's approval to invite tenders. They should discuss and agree the status of the design team during the contract period and whether the client's design team will be novated to the contractor or if the contractor will use their own design team, so that the client's design team can monitor the works on the client's behalf.

For a one-stage design and build project, the tender documents will include the Employer's Requirements. These will be similar to those issued at Stage 2 for a two-stage design and build but will have been developed in much greater detail.

As with other procurement routes, the contract administrator will assemble and issue the various tender documents and arrange for the receipt of the tenders. Once a preferred contractor has been selected the contract administrator should carry out the pre-construction tasks as detailed for Stage 4 (see page 118) to enable the works to commence on site at the beginning of Stage 5.

The Employer's Requirements for a single-stage project will be more comprehensive and detailed than for a two-stage project, reflecting the greater detail to which the scheme will have been developed by the client's design team to enable the client to have a firm tender price immediately.

The Employer's Requirements should include the following items (the list is not exhaustive):

- preliminaries and general conditions
- site details
- site constraints
- topographical surveys
- geotechnical reports
- planning permission and other consents together with conditions, if known
- any statutory consultations held
- existing health and safety files, if appropriate

- functionality of the building
- architectural drawings
- engineering drawings
- BIM model
- project strategies
- other designers' information
- schematic layouts/flow charts
- schedules of accommodation
- room data sheets
- specification information
- equipment requirements
- programme information
- schedules of work
- schedules of rates
- details of the contract
- preliminaries, insurances, novation, etc.
- information to be included in the Contractor's Proposals
- content and form for the contract sum analysis.

Following the preparation of the tender documents, the contract administrator should follow the general rules and principles for the selection and tendering process as detailed for Stage 4 (see page 93).

The tender return documentation will take the form of the 'Contractor's Proposals'. These show how a contractor bidding for the project intends to fulfil the Employer's Requirements. They set out the contractor's design and how the contractor proposes to construct the building, together with their price.

The format and content of the Contractor's Proposals should be as described in the Employer's Requirements, but they might include:

- detailed design drawings prepared by the contractor's design team (novated or otherwise)
- trade specifications the contractor has developed during the tender process
- schedules
- method statements on how they propose to carry out discrete parts of the works
- a draft Construction Programme

- details of inconsistencies between the Contractor's Proposals and the Employer's Requirements
- a list of proposed subcontractors
- details of any proposed provisional sums
- the tender price and a 'contract sum analysis'.

The 'contract sum analysis' is the priced document prepared by the contractor as part of their tender. It breaks down the contractor's price into the format stated in the Employer's Requirements which allows the client and contract administrator to analyse it and to compare it with the other tenders. The format should include preliminaries and an elemental breakdown of trades. Once construction starts, the contract sum analysis can then be used by the contract administrator as the basis for calculating payments due to the contractor as the works progress.

The contract administrator should evaluate each section of the proposals, not only for price, but also for the design proposals, method statements and programme together with the contractor's team. There is likely to be a period of negotiation to check for, and agree, any inconsistencies between the Contractor's Proposals and the Employer's Requirements. This is a very important part of the tender analysis as it will ensure that it is always completely clear which document prevails after the contract has been entered into.

As with other collaborative style contracts, where it is important that the teams work well together, the shortlist of prospective contractors should be invited for interview before a final decision is made. This gives the contractors the opportunity to expand on their proposals, ask questions, answer queries and introduce their project teams.

At the end of Stage 3 the Building Contract will be awarded to the selected contractor. The contract administrator should confirm this in writing as a letter of intent and prepare the contract documents for signature by the end of the stage, so that the detailed design work can start immediately at the beginning of Stage 4.

Contractor-led contract

The contract administrator will have issued the details of the tender return and the format for the Contractor's Proposals with the tender documents at the end of Stage 1. During Stage 3, only the preferred bidder selected during Stage 2, will develop their concept design into

Contractor's Proposals, which will be used as the basis for award of the Building Contract at the end of the stage.

The contract administrator should monitor and instigate any Change Control Procedures during this stage (see figure 4.15 for a suitable format). On a contractor-led project, change control is more contractually focused, but it is most important to get the change control culture in place at the start of Stage 3.

Following the receipt of the Contractor's Proposals, the contract administrator will evaluate the tender and report to the client prior to the award of the contract.

On the basis of the Contractor's Proposals being acceptable, the contract administrator should then prepare the contract documents for signature by the end of the stage, so that the detailed design work can start immediately at the beginning of Stage 4 (see page 122).

Chapter summary 3

This chapter has described tendering activities for one-stage design and build and contractor-led projects during Stage 3. When tenders are submitted at this stage, the design will contain more detail than if the project were tendered at an earlier stage.

A key role of the contract administrator is to ensure that the client will obtain competitive tenders in terms of both cost and design by following the rules and procedures in a methodical way.

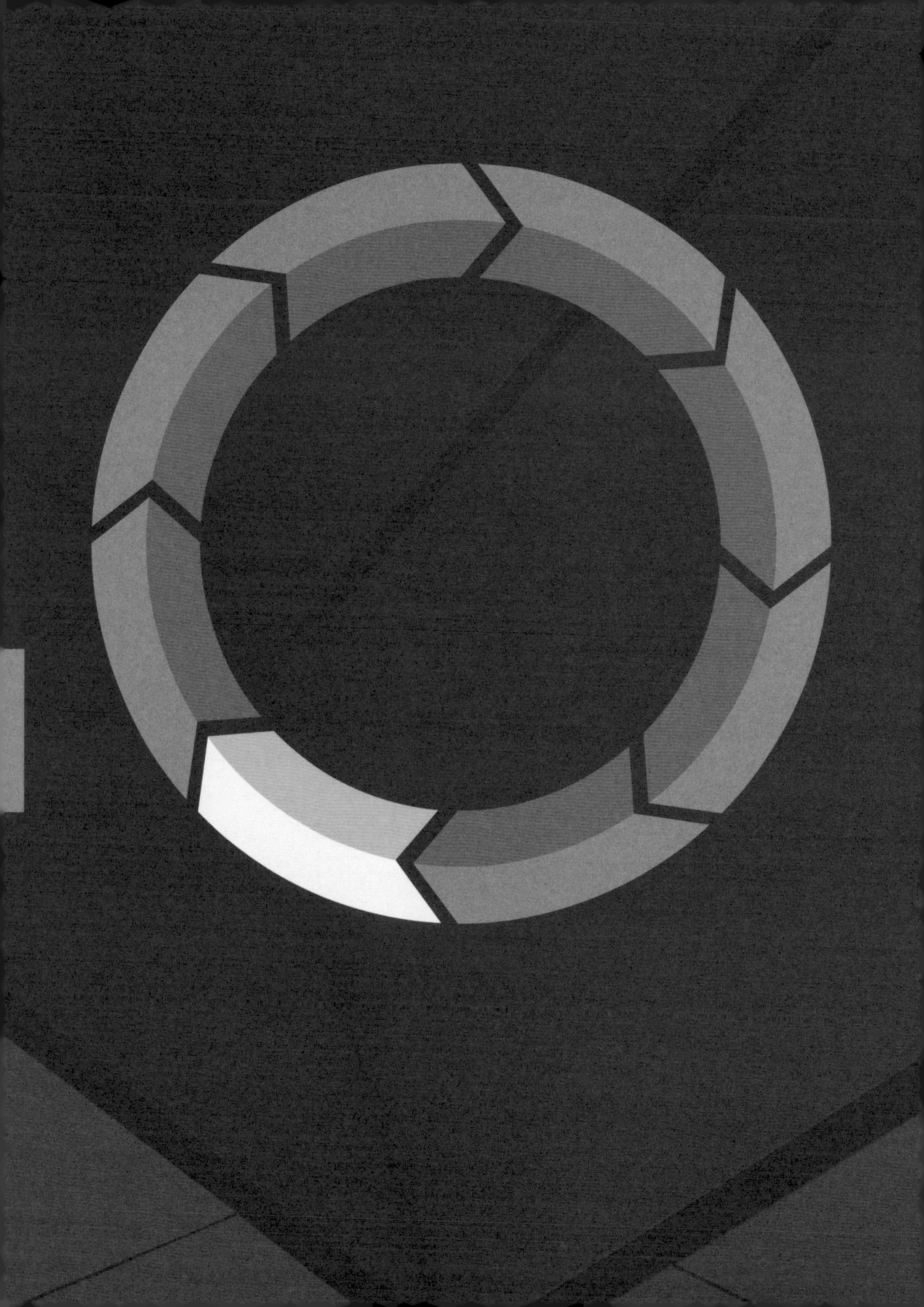

Stage 4

Technical Design

Chapter overview

During Stage 4 the design team will be completing the technical design in sufficient detail to invite tenders on traditional projects and conclude the contract on two-stage design and build forms of tender.

This chapter considers in more detail how the contract administrator takes the information prepared by the design team and assembles it with other documents to enable competitive prices to be obtained from selected contractors. It considers the actions that the contract administrator takes in the preparation of the tender documents, the forms and procedures for inviting and receiving tenders applicable to a traditional contract, together with work associated with the preparation and signing of the contract documents immediately prior to operations starting on site. The tasks and procedures described are applicable to tendering for alternative procurement routes with appropriate adaptations in the relevant wording.

The key coverage in this chapter is as follows:

What tendering activity is there at Stage 4

What is tendering?

What is the tender process?

Preparation of the contract documents

What about change control?

Introduction

The priorities for the project team, and particularly for the contract administrator, during this stage will be to organise and manage the successful receipt of competitive prices for the works so that a decision can be made on which contractor should build out the project. At an early stage the contract administrator should describe the tender process to the client so that the client understands the implications for the project. The client should also be made fully aware that if they make any changes to the design after the tenders are returned, those decisions will result in variations that may add to the construction cost as well as increase fees and cause delays to the works.

The tasks and procedures involved in the tender process are very similar across the various procurement routes, although they vary according to the level of detail and the stage at which the events occur. This chapter deals primarily with the tasks and procedures for a traditional form of contract, but they can be simply adapted to suit whichever form of contract is selected.

It is beyond the scope of this guide to examine every tender variation or procurement route in detail. Instead, it aims to act as a quick reference guide or aide memoire, covering the main options and issues to look out for. While the principles of tendering are discussed, for more detail on tendering reference should be made to the JCT's *Tendering Practice Note 2012* or the *NBS Guide to tendering: for construction projects* (www.thenbs.com) or other similar sources appropriate to the chosen form of contract. For public contracts, reference should be made to The Public Contracts Regulations 2006 and The Public Contracts (Amendment) Regulations 2011.

What are the Core Objectives of this stage?

The Core Objectives of the RIBA Plan of Work 2013 at Stage 4 are:

The Core Objective at Stage 4 is for the design team to complete the design in sufficient detail (as appropriate to the procurement route) to enable construction work to commence at the beginning of Stage 5. Although competitive tenders will have been obtained at earlier stages for other forms of contract, tenders on traditional contracts will be invited during this stage, leading to the awarding of the Building Contract at the end of Stage 4.

What supporting tasks should be undertaken during Stage 4?

The Suggested Key Support Tasks noted in the RIBA Plan of Work 2013 have been devised to support the Core Objectives and to ensure that the documentation required to proceed to the next stage has been prepared. At this stage, the Suggested Key Support Tasks for the contract administrator are:

- Review and update Maintenance and Operational and Handover Strategies.
- Prepare and submit Building Regulations submission and any other third party submissions requiring consent if not undertaken by the design team.
- Review and update Project Execution Plan.
- Review Construction Strategy, including sequencing.

Within a bespoke Plan of Work based on a traditional project (see figure 4.1) the variable procurement task bar fixes the support tasks for the contract administrator at Stage 4 relating to tendering. This generates key outputs which will include finalising tender documents, for which any of the following documents and information could be relevant:

- a list of all tender documents, so that the tenderers can check they have received the complete package
- tender forms and details of the procedure to be followed, eg type of tender required, submittals required, how the tender should be packaged and identified, to whom it should be sent
- whether there is any contractor's designed portion
- Site Information and surveys
- drawings and/or BIM model
- drawn schedules, eg for doors and windows
- specification
- bills of quantities
- schedule of works
- schedule of rates
- activity schedule
- the health and safety plan
- programmed dates for proposed work
- details of any phased commencement or completion

- details of the contract terms and conditions, including insurance provisions
- details of any bonds or guarantees required from the contractor or to be provided by the client
- details of any warranties to be provided
- information prepared specially for use in self-build or semi-skilled operations.

What procurement activity is there at Stage 4?

Traditional forms of contract will be tendered at this stage. Building Contracts will have already been awarded for one-stage design and build, management and contractor-led projects, so there is no formal action required by the contract administrator at this stage.

Stage 4 actions for a one-stage design and build relate to the contractor's design team and specialist subcontractors developing the detail design in accordance with the Design Programme.

For two-stage design and build at Stage 4 the contractor will be working with the design team to assist in the development of the design to produce the technical information that will comprise the Contractor's Proposals. If the price is satisfactory, the Building Contract is awarded during the stage, although some further design work may be undertaken by the design team and specialist subcontractors.

On management contracts the detailed design information based on agreed work packages will be developed and separately tendered during Stage 4 to allow sufficient lead-in time to meet the contractor's Construction Programme on site. In this instance there is likely to be an overlap between the design and construction periods at Stages 4 and 5.

No further procurement activity takes place on contractor-led projects at Stage 4 other than the contractor tendering for its labour and materials.

What tendering activity is there at Stage 4?

Tendering activities undertaken by the contract administrator at Stage 4 solely relate to the traditional form of tender.

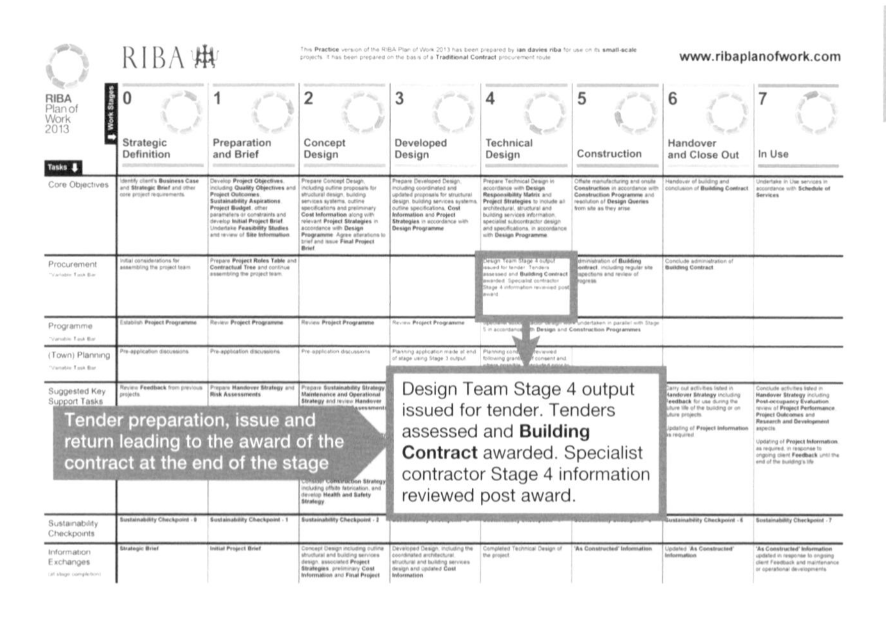

Figure 4.1 Bespoke RIBA Plan of Work 2013 for a traditional project

The detailed design work for a fully detailed scheme will be complete during Stage 4 to enable the contract administrator to prepare the tender documentation, invite tenders and report on them leading to the award of the contract to the successful contractor at the end of the stage.

The contract administrator should follow the general rules and principles for the tendering and contract award process as detailed below.

What is tendering?

Tendering is the method used to select the contractor for a building project. Contractors who are interested in carrying out the construction work are invited to submit compliant bids for that work – the best bid is then chosen in a fair and transparent way. The objective is to obtain the best value for the project from the contractors who are best able to carry out the works to the timescale and quality required.

The principles of tendering are the same whatever the procurement route. Tenders can be invited on the basis of the lowest price or the most economically advantageous option (which is common in public sector and EU procurement). In January 2014, new rules were agreed by the European Parliament that introduced a new criterion of the 'most economically advantageous tender' (MEAT) in the award process.

The contract administrator should not make the tender documentation overcomplicated and should consider the time and cost involved for the contractor in preparing a tender.

On what basis can you invite tenders?

There are three main forms of tendering: open, selective and negotiated. The option used will determine how many contractors are asked to submit tenders.

- *Open* tendering is where all contractors who have pre-qualified following an advertised invitation to tender are asked to submit a bid. This is very competitive but is wasteful of resources, although it is not uncommon within EU procurement.
- *Selective* tendering is by invitation only – only a predetermined number of contractors are invited to bid, the number depending on the type of project. This is appropriate for all types of procurement and project types and is fair and clear to all parties. The selection of tenderers will be based on pre-qualification against specific criteria and previous experience.
- *Negotiated* tenders are commonly used in follow-on projects, where the client knows the contractor and trusts them to give a fair price – costs can be based on the previous project with a percentage uplift. This form does not necessarily achieve the most competitive price except where it forms the second part of a two-stage tendering process and the management costs and rates have been tendered competitively at the first stage.

Paper or electronic?

Traditionally, tender documents were issued to the contractors in paper form and the tenders were returned by post or by hand. However, with the increased use of documents in a digital format it is increasingly likely that tenders will be issued and received electronically. The procedures which follow are based on a traditional paper format, but they can quite easily be adapted into a simple electronic format for use on small projects, or, for larger, more complex schemes, in a common data environment.

Electronic tendering (e-tendering)

E-tendering is the electronic issue and receipt of tenders, which simplifies the process and reduces costs. Electronic documents can be exchanged in a number of ways, but steps must be taken to maintain confidentiality and ensure that documentation is secure, to avoid it being changed or going astray. The use of digital signatures and certificates enhances the security.

From the outset the software to be used and the file formats (Word, Excel, CAD, etc.) should be agreed, together with the version number. For example, a company may not be using the latest version of their CAD software so all drawings would need to be issued in the oldest format used by all of the tenderers. Text-based documents should be page referenced and drawings should be issued in a format where they can be printed to the correct scale and line format. If CAD drawings are issued then the plot files should be included to ensure correct printing. Information can be exchanged in a number of ways:

- *On disk:* DVD, CD or memory sticks are familiar electronic formats usually sent in the post with the tender documents similar to paper copies. However, it is expensive and time consuming to reissue the documentation in this format.
- *By email*: This is a more informal communication medium. It is most important, when issuing tender documents by email that this is followed up with each contractor to make sure that they have received them in case the email has been rejected because the file size cannot be accepted by the server. Although email is useful for quick exchanges and the issue of supplementary documents, questions and queries, there are potential security issues which are against the traditional principles of sealed bids and confidentiality.
- *Websites:* As well as using public sites, such as Dropbox and Google, companies will often create a common data environment (CDE) by the use of extranets. An extranet is a computer network that is accessible by specific people or organisations to store design data or tender information. Each tenderer will have their own secure area in which to upload their tender and large volumes of data can be exchanged efficiently in this way

Electronic tendering (e-tendering) (*continued*)

and the use of encryption also makes the process more secure than sending data over the public internet.

The use of e-tendering should not affect the normal tender timescales, the procedures involved or the adopted standards for assessment. However, control of the documentation is equally important as with paper copies, if not more so.

Document control

Whether using paper or digital formats every document should be assembled, referenced and put in appropriate, clearly labelled folders, with each document having its own file with a unique file name. Good file naming practice is imperative – each file should have the job number, the name of the document, the revision number and the date for auditing and tracking, eg: **5210 Specification v3 27apr14.**

What is the tender process?

To ensure fairness and clarity, tendering procedures are governed by best practice, and EU public procurement rules and regulations apply for contracts valued above certain thresholds. This section sets out a traditional method of selecting and inviting tenders suitable for either a manual or electronic system, with the same principles and forms of words applying in either case.

The contract administrator should be confident that each contractor chosen to tender will be able to undertake the work. To achieve this, a series of eligibility criteria should be set for the project, either by the client and design team or by statute, against which prospective contractors are assessed to determine their ability to complete the project. The criteria, timetable and procedures for public sector contracts are set by EU directives.

What is pre-qualification?

Contractors are usually pre-qualified before being invited to tender. The purpose of prequalification is to prove the capability of a contractor before they are invited to tender. It is usual to prepare a list of pre-qualified contractors so that any one on the list could do the job, thus making the final choice of contractor after the receipt of tenders a simple one.

Selection criteria should be set for contractors based on their competence, financial standing, capability, experience and references. Pre-qualification questionnaires can be used to simplify the process of selection, although these are often overcomplicated.

The pre-qualification list can be prepared in a number of ways, depending on the size and nature of the project:

- The client may have their own list of preferred contractors with whom they have worked before.
- The list may be prepared by the design team based on past experience of working with the contractors.
- Lists of approved contractors may be used – good for repeat or similar projects.
- Contractors may be invited to complete a pre-qualification questionnaire (PQQ).

Assembling the tender list

Sometimes the contract administrator may have to assemble a tender list in an area where they do not know any suitable contractors. On larger projects it is likely that national building contractors with whom the contract administrator has worked in other regions will have local or regional offices. In this case it is a simple task to meet with the local representatives and interview them. With smaller projects and those with private clients it is more difficult to identify potential contractors. This will need good local knowledge, which could be obtained through the local branch of a professional body such as the RIBA and RICS or by directly contacting local practices who work on similar projects.

- Notices may be placed in the *Official Journal of the European Union* (*OJEU*) for projects with contract values over the EU threshold. Contractors are shortlisted based on the satisfactory completion of a PQQ.

The simplest and most flexible selection process is to list known contractors based on their reputation and past experience. However, even with known contractors and when using approved lists, the list should be reviewed periodically to ensure that each contractor's details, workload, availability and references are up to date. New companies should be added as necessary – they may come to attention through their reputation or they may write in to the practice. Indeed, there are always good contractors in the market who the practice may have not heard of but who are more than capable of doing the work. These should be checked out and records made based on their competence, financial standing, capability, experience and references.

The client's contractor?

Be cautious about adding to the tender list any contractors the client suggests, particularly on smaller projects, without fully checking out their capabilities and competence. If a recommended contractor is then appointed on the basis of being the cheapest, there is a risk that the contract administrator could end up with much more work and a client who is not happy about the quality and time taken to complete the project. Invariably it will be the contract administrator who is deemed to be at fault, not the contractor! It is better to check out the contractor and advise the client against using them or, if the client is insistent, express any concerns to the client in writing before appointing 'the client's' contractor. That said, experience has shown that using the client's preferred contractor is not necessarily a bad thing – in some cases you end up finding a contractor who you would be happy to use on future projects.

What are pre-qualification questionnaires?

PQQs are intended to simplify the process of contractor selection. However, this is not always the result – they can be very time consuming for contractors (and indeed consultants) to fill in and there is no guarantee of progressing to the next stage.

PAS 91:2013, which was sponsored by the Department for Business, Innovation and Skills (BIS), is a standardised PQQ created to harmonise the various and numerous question sets used between different buyers to make PQQs simpler for the supplier to complete.

PQQs set the basic eligibility criteria and reduce the likelihood of non-compliant tenders being submitted by under (or over) qualified and under-resourced companies. Getting basic checks out of the way at this stage means that the subsequent tender documentation can be wholly project specific.

The questionnaires should be as simple as possible, with the basic questions limited to determining the organisation's:

- structure
- financial capability
- technical capability
- current workload and capacity
- references
- experience on recent/similar projects
- health and safety resources/safety history
- quality assurance accreditations
- environmental policy.

Questionnaires should be returned by the date set for their return and evaluated on the basis of the criteria set for the project. Records should be made of the evaluation process and references taken up as necessary.

How are the tenderers selected?

Tenderers should be selected on the basis of equal standing, suitability and experience. As well as having a shortlist it is always prudent to select two reserves.

The tender list should be kept to a reasonable size, so that each contractor's chance of winning is balanced against their potentially abortive costs and the time required to prepare their tender.

Having established which contractors are best suited for the project and prepared the initial list of possible tenderers and before issuing a formal enquiry letter, it is well worth checking if they are still interested in tendering for the project. A direct telephone call will save time and the contractor's level of interest can be gauged at the same time. At this point the contractor may have to decline as they no longer have the capacity to do either the tendering or the contract work. It is better that the contractor declines on the basis that it would not prejudice them from tendering for future projects, rather than the contractor saying they will tender and backing out after tenders have been issued or putting in a 'cover price' which they know is too high. Historically, a cover price has been used when a tenderer is too busy but does not want to offend the client by not tendering, this is unsatisfactory to all parties.

Preliminary enquiry

After informally checking which contractors are interested, a formal preliminary enquiry letter should be sent to each prospective tenderer outlining the details of the project and the tendering details. The tenderers should be advised if electronic tendering will be used and ask for confirmation of preferred file formats.

An example letter is shown in figure 4.2, but an alternative is to set out the same detailed information as a Project Information schedule, which would be accompanied by a simple covering letter.

The letter should give the date by which the contractor should confirm their interest. The timescale will vary according to the type and complexity of the project, but it is likely to be up to two weeks for a small project or 14–21 days for a larger one. For public sector procurement the period is longer, more likely 37 days, or 30 days if transmitted electronically. However, due consideration should be taken where a bank holiday occurs over the period or during the school holidays.

Tender lists for small projects

On small projects, particularly ones that are similar to previous projects, it is likely that the potential contractors are already known to you. In such cases, an initial round of telephone conversations is all that is needed to compile the shortlist. That said, the principles behind the information imparted and the evaluation of contractors are the same as for larger projects, albeit less formal.

Example preliminary invitation to tender letter

Dear

Re: PROJECT TITLE

Further to our telephone conversation I confirm your interest in tendering for the above project and we are now preparing the list of tenderers. So that we can finalise this we would be grateful if you could formally confirm whether you wish to submit a tender on the basis of the following project information if invited to do so.

Project title:

Employer:

Architect/Contract Administrator:

Consultants:

Location of site: (include site plan)

General description of the work:

Form of Contract: (insert form with supplements as required eg with contractor's design)

Contract to be executed: (as a deed/by hand)

Contractor's design responsibility: (where the contractor's design option applies)

Fluctuations:

Insurance options:

Liquidated and ascertained damages: (£ per day/week)

Bond/guarantee/warranty requirements:

CDM: (applies/does not apply)

Health and safety planning period:

Approximate cost range: £ to £

Approximate value: £

Anticipated start date:

(continued)

Figure 4.2 Example preliminary invitation to tender letter

Example preliminary invitation to tender letter (*continued*)

Contract period:

Approximate date for issue of tenders:

Tender period: xxx weeks

Date/time for receipt of tenders:

Proposed number of tenderers:

Tender to remain open for xxx weeks:

Method of submitting tenders: electronic/hard copy

Requirements for alternative tenders:

Tender assessment basis: lowest price/best value

Any other significant information:

Tenders will be invited on a single-stage/two-stage/negotiated basis and If you wish to be invited you must agree to submit a bona fide tender in accordance with the relevant published procedural notes and guidance on the selected form of tendering. To ensure fairness and maintain confidentiality you must not divulge your tender price to any party before the tender due time. When the contract has been signed you will be advised of the names of the tenderers and the prices received.

Should you no longer be able to tender then please advise us as soon as possible but this will not preclude you from being invited for future projects.

If you wish to be invited then please confirm this in writing no later than .... 20.., but please do not hesitate to contact us should you have any queries or need further information.

Yours sincerely

Final selection

After the receipt of responses to the preliminary enquiry, a final list can be prepared and references taken up, if not already requested. The list should then be agreed and confirmed in writing to the client.

At that point the successful tenderers should be advised that they have been included on the tender list and when to expect the issue of the tenders, so that they can have sufficient resources available. If there are any shortlisted but unsuccessful candidates, they should be advised that they are not being invited in this instance but that they are on the reserve list.

What should be included in the tender documentation?

The documents issued with the tender list normally comprise variations of the following items, depending on the form of procurement being used.

Preliminaries and general conditions

These describe the works in a clear and precise way, and specify general conditions and requirements, such as contract details, time constraints, approvals, testing and completion. This will allow the contractor to assess costs that, while not forming part of the direct construction cost, need to be priced as a result of the chosen method of construction.

These costs may either be one-off fixed costs, such as the cost of bringing to site and erecting site accommodation (and subsequent removal), or time-related costs, such as the heating, lighting and maintenance costs for that accommodation, scaffolding, skips and so on.

There should also be allowances for provisional sums for specialist items such as alarms, fittings, etc., together with a contingency sum to cover any unforeseen matters. Generally speaking, the contingency should be in the order of 5% on new work and 10% on alteration works.

Setting up the preliminaries

The preliminaries and general conditions for a range of JCT contract forms can be set up using *NBS Building* or *NBS Create*. Produced by NBS, these are based on the National Building Specification, which is in turn based on the classification document Common Arrangement of Work Sections for Building Works (CAWS) published by the Construction Project Information Committee. This covers general preliminaries items, such as the project details through to more detailed requirements on safety, security, sequence and timing of the work, facilities and services and so on. There is also provision for allowances for statutory services, provisional sums and contingences.

Design team drawings

Drawings can either be fully detailed for a traditional format or to the level of detail required for a one- or two-stage design and build project.

These may cover architectural and structural design, services design, etc. provided by consultants, depending on the size and complexity of the project.

BIM model

This includes a 3-D digital representation of the building using Building Information Modelling (BIM), a process involving the generation and management of data to create files which can be exchanged and worked on by the design team. This will have a greater impact on tendering in the future should BIM models be made available to tendering contractors to enable them to understand the complexities of the project in more detail. Further on, as BIM begins to extract quantities from models, it will provide more detailed and accurate documents for tenderers to price.

Health and safety information

This information is usually included as part of the preliminaries and general conditions, but where a project is notifiable under the Construction (Design and Management) Regulations 2007 (CDM regulations) and a health and safety adviser has been appointed, pre-construction information should be prepared for issue with the tender. Pre-construction information is intended to ensure that health and safety issues and essential resources are properly considered to minimise the risk of harm to those who build, use and maintain the building.

Health and safety information

Further reference should be made to *Managing Health and Safety in Construction: Construction (Design and Management) Regulations 2007 Approved Code of Practice (HSE Books, 2007).* Appendix 2 refers to information which may be required for the pre-construction information.

Specifications

Compiled by all members of the design team, specifications – whether descriptive or performance based – define the materials to be used, the quality of work to be achieved and any performance requirements or conditions under which the work is to be executed. Descriptive specifications state the actual products to be used, such as the type of insulation, whereas performance specifications state the technical performance criteria to be achieved, such as the K value, rather than the material itself.

The Specification should be related to the 'level of detail' agreed within the Design Responsibility Matrix, whether performance or full (generic or proprietary).

Specifications should be clearly defined as any omissions or conflicts could cause delay or extra costs when the project gets on site.

Compiling specifications

Specifications can be set up to suit the particular form of contract being used (eg JCT Standard, Intermediate or Minor Works or variants of these) by using *NBS Building* or *NBS Create* (produced by the NBS) or other similar systems.

Bills of quantities

Some project types require bills of quantities to be prepared, usually by the cost consultant. Bills of quantities are becoming less common, but they are the traditional way of obtaining tenders for a project where the design is fully detailed at tender stage, enabling contractors to price on the same basis.

Bills of quantities should be prepared according to an industry standard to avoid any ambiguities or misunderstandings. This will help to avoid disputes arising from different interpretations of what has been priced, as could happen if each tenderer measures and prices their own bill from the drawings and specification. As an alternative to a full bill of quantities, on projects where the design is not yet complete, an approximate bill of quantities can be used for the tender. However, this will inevitably lead to more variations and the actual works will need to be re-measured on completion, so there is less price certainty at the time of tender.

Schedules of work

Schedules of work are generally prepared by the designers, rather than the cost consultant, and should simply list the work required. The work could be arranged on an elemental basis (eg brickwork, plastering, decorations) or on a room-by-room basis. A schedule can also include a description of the work required, which is known as a 'specified' schedule of work.

Schedules should allow the contractor to identify all the work and materials needed to undertake the works and to calculate the quantities required from the drawings. As a consequence, it is important that schedules of work properly describe every significant item of work to which they relate. Otherwise, the tenderer may miss out an element of the works, which could result in a claim by the contractor at a later date.

Any other items

As a catch-all you can always add an additional section at the end of a schedule of work to put more onus on the tenderer to check the drawings and the specification for other items, so that nothing is omitted from their price.

20.5	*Allow to leave the site neat and tidy on completion*	
21.0	***ANY OTHER ITEMS***	
21.1	*It is assumed that all items noted with the drawings and specifications are included in the tender. Therefore, if you have any other items you wish to allow for, which may be on the drawings/specification but are not noted in the schedule, you should list them below.* • •	

However, this clause should not be seen as a substitute for fully considering all the elements required and ensuring they are included the schedule.

When pricing a schedule of work the tenderers should be advised that they should price each item individually and not group them together. Although it is often simpler for the tenderer to group items where they have received a subcontract price or they have priced the job section rather than each element, it has been found in practice that where the contractor prices each element individually, the total price will be higher. On smaller projects contractors tend to price the job as a whole and then (if successful) break down the overall cost into the scheduled items.

Schedules of rates

These are sometimes used as the basis of selecting the preferred tenderer where there is insufficient detail available at the time of tender and so the contractor is unable to establish the quantities required or in the first stage of a two-stage tender process.

In its simplest form a schedule of rates can be a list of staff costs, types of labour and plant against which a contractor prices specific rates for all activities that might form part of the works. Indicative quantities may or may not be given to tenderers, but they do not form part of the contracted works. General preliminaries such as scaffolding, temporary power, supervision and temporary accommodation will also be priced.

The schedule of rates can also be used on traditional contracts to estimate the cost of variations or to calculate the cost of any dayworks instructed by the contract administrator.

How are the tender documents prepared?

The contract administrator should ensure that a full record is kept of all the documents and their status or revision, whether they be incoming or outgoing. Whether this is done online or manually, the principles and the information to be recorded remain the same.

On larger projects, drawings and documentation are now commonly produced in a 'common data environment', where all of the Project Information is stored on a dedicated website in the 'cloud'. This ensures not only that the information is secure, but also that, by careful management, all members of the team are working to the correct information concurrently, thus minimising the possibility of working with out-of-date information.

Traditionally, and on the vast majority of smaller projects, the contract administrator monitors the issue of drawings at various stages (whether for planning, tender, construction etc.) via a drawing register/drawings issue sheet, such as that shown in figure 4.3. Each consultant should prepare and use their own list. Setting these up at the start of the project provides a full audit record of the issue of the document revisions through to completion.

Issue sheets can also be used to record the issue of other project documents, such as the specification and reports. However, it is equally

Drawing issue sheet

Ed Associates Ltd.

141 New Business Park, Anytown, AT51 1ST

JOB No.	JOB TITLE
123	**New Music Centre**

Drawing Issue Sheet Sheet 1 of 1

No.	Title	Scale		Date	
				26	26
				4	4
				13	13
01	Location Plan	1:1250	A4	x	x
02	Block Plan	1:500	A3	x	x
03	Ground Floor Plan	1:100	A1	x	x
04	First Floor Plan	1:100	A1	x	A
05	Roof Plan	1:100	A1	x	A
06	Elevations	1:100	A1	x	A
07	Sections	1:20	A1	x	x
08	Details	1:10	A1	x	x
09	Doors/Windows	1:20	A1	x	x
10	External Works	1:100	A1	x	x
Format	P=Paper E=Email D=Disk F=Fax			E	P
Status	PR=Preliminary PL = Planning A=Approval T=Tender C=Construction AB=As Built			A	T
Distribution					
Client		Contact info.		1	1
Structural				1	1
Landscape				1	1
Planning					
Bldg Control				2	
Contractor/tenderer					2
File				1	1

Figure 4.3 Example of a drawing issue sheet

Revisions to drawings

Everyone involved in a project must ensure that they are using the current versions of the documents, both at pre-contract stages and during construction. The drawings register/issue sheet should be updated constantly, and reissued to all affected parties. At the end of the project this should become the register of as-built drawings.

CAD systems make revising drawings easier than by hand and there will always be the temptation to reissue drawings in response to the smallest site queries and design changes. However, it is important to establish a thorough Change Control Procedure to manage this, so that revisions are made, and issued, only when appropriate. When any revision is done, either a paper or electronic copy should be made of the previous revision so that a full record will be available at a later date.

Whether the drawings are being revised by hand or on CAD, the revisions notes on a drawing must be sufficiently specific, eg not 'General revisions'. CAD software enables revisions to be highlighted with 'clouds' and text on a separate layer, and these should be used to identify parts of the drawing that have changed. The revision reference should be attached to the cloud, and the text in the revision box should give a concise description of the revisions carried out. This system can equally be applied to hand drawings by putting the clouds in pencil on the back of the negative, making it simple to erase the old clouds at the next revision.

important to record all *incoming* documents and to note the actions required on them. An example of this is shown in figure 4.4.

The preliminaries and contract details should be finalised with the client. Once the preliminaries are drafted a meeting should be arranged with the client to confirm the details, such as:

- access
- client constraints
- contractor's working area
- contact for site visits
- start and completion dates

Document received sheet

Ed Associates Ltd.
141 New Business Park, Anytown, AT51 1ST

Sheet 1 of 1

JOB No.	JOB TITLE							
123	**New Music Centre**							
	Record of documents RECEIVED							
Doc. Ref	**Drawing/Report/document title**	**Doc. date**	**Revision**	**From**	**Date Rec'd**	**Reply by**	**Reply date**	**Notes**
A4016	*Site Investigation Report*	*Mar-13*	*A*	*Groundworks Ltd*	*26-Mar-13*	*N/A*		*Extra trial hole added*
5425/006	*Structural plan*	*24-Mar-13*		*Oldem Upwell Ltd*	*27-Mar-13*	*03-Apr-13*	*01-Apr-13*	*Initial proposals*
685/001	*Acoustic Report*	*27-Apr-13*		*Hearwell Acoustics Ltd*	*31-Mar-13*	*07-Apr-13*		

Figure 4.4 Example of a list of documents received.

- amount of liquidated and ascertained damages
- insurances
- guarantee bonds and warranties
- limitations on working hours and access
- payment details and retention percentages
- health and safety issues
- advance appointments and works by the client or others employed directly by the client
- sections of the works that the contractor needs to design (the contractor's designed portion).

From experience, assuming that the full CDM regulations apply, it is useful if this meeting is also attended by the health and safety adviser to enable them to address their queries prior to finalising the pre-construction information for inclusion in the tender documentation.

As well as the tender issue documents the tender return documents will need to be prepared. The prime document is the 'form of tender', but the tenderers should also return a summary breakdown of their costs. This will enable the contract administrator to make a quick comparison, without having to investigate any detailed breakdowns that may accompany the tender.

A tender summary, as indicated in figure 4.5, should be set out in the same way as the sections in the specification document, with the preliminaries, specification and schedule of work as the main sections and then subdivided as necessary.

The tender form will not only indicate the overall tender figure for comparison, but should also confirm the tenderer's lead-in period, actual/earliest start date and contract period. This will be in a form similar to figure 4.6 but adapted to suit the particular procurement route selected and form of contract.

If the tenders are to be submitted in the traditional return envelopes then envelopes or labels will need to be prepared. These should simply have the return address, job number, the name of the project and the time and date of return on it.

There should be no other names or markings on the return envelope. However, it is always useful when preparing the tender list to number

Tender summary form

NEW MUSIC CENTRE

AT AGOOD SCHOOL

LONDON ROAD, ANYTOWN, AT1 1TA

TENDER SUMMARY

	£	p
SECTION A – PRELIMINARIES & GENERAL CONDITIONS	£	p
Total brought forward from Section A		
(including Provisional Sums A54/A55)		
Contingencies	5000.00	
SECTION B – SPECIFICATION	£	p
Total brought forward from Section B		
SECTION C – SCHEDULE OF WORKS	£	p
Totals brought forward from Section C		
PART 1 – DEMOLITIONS		
PART 2 – NEW BUILD WORKS		
PART 3 – ALTERATIONS TO EXISTING		
PART 4 – EXTERNAL WORKS		
TOTAL CARRIED TO FORM OF TENDER	£	p

Date

Figure 4.5 Example of a tender summary form

the tenderers and put those numbers on the labels (eg 2/job number). The contract administrator will then know which tenders have not been returned at any point in the tender return process.

In practice, it is more efficient and cost effective to use address labels rather than return envelopes. Each contractor can then choose the optimum size of envelope and packaging to suit their submission. This

Form of tender

NEW MUSIC CENTRE

AT AGOOD SCHOOL

LONDON ROAD, ANYTOWN, AT1 1TA

FORM OF TENDER

We, the undersigned, do hereby tender to execute, complete and maintain the work set forth and described for the above-mentioned Contract, in accordance with the Conditions of Contract and Specification and Drawings prepared by Ed Associates Ltd, and to their entire satisfaction, for the sum of:

....

.... £______________:______

We confirm that we would be able to commence the Works described in the tender documents within .... weeks from acceptance of our tender with a .... weeks contract period. *

or * on__________ for completion by____________________(......... weeks)

I/We estimate that the Employer's Liability for Value Added Tax in respect of the Works will be

£__________

Dated this .... day of .... 20......

SIGNATURE ....

ADDRESS ....

....

This Tender is to remain open for acceptance for a period of thirteen weeks from the date of Tender.

THE TENDER SUMMARY MUST BE COMPLETED AT TIME OF TENDER AND RETURNED, TOGETHER WITH FORM OF TENDER.

Alternatives*

Figure 4.6 Example of a form of tender

is particularly helpful with design and build contracts where tenders are likely to include drawings and specifications in addition to the Contractor's Proposals.

How is the tender issued?

Once the tender documents are ready for issue, the contract administrator should obtain written confirmation from the client that they want to invite tenders. The document packs should be issued to the tenderers with guidance on how the documents should be completed and the return details. On traditional contracts there should be between three and six tenderers, and four on design and build contracts to reflect the more detailed design input required by the contractors during the tender period.

The contract administrator should ensure that tenderers are allowed sufficient time to prepare their tenders. If tenderers have insufficient time to make enquiries and obtain specialist quotations it is likely that their tenders will be higher than they would be otherwise. This is quite often the case with mechanical and electrical subcontracts, which often come in at the last minute (or late), resulting in the tenderer asking for an extension of the tender period.

It is considered that three to four weeks is necessary on a small and simple traditional project, but longer will be required for a design and build project.

A formal invitation to tender letter should accompany the tender documents, summarising the nature of the tender, what documents are enclosed and the details for the return of tenders. An example letter is shown at figure 4.7, but the same information should be included if the information is issued electronically.

It should be noted that the letter includes a statement to say that '*the Employer reserves the right to postpone the date for the return of tenders and it is to be understood that neither the lowest nor any tender will necessarily be accepted. No tender expenses will be payable*'. This should also be included in the specification or bills of quantities to safeguard the client from any claims being made against them by a contractor whose tender is declined or if the lowest tender is not the one accepted.

Tender invitation letter

Dear Sirs *– other options*

Re: PROJECT TITLE

Further to your confirmation that you wish to tender for these works we enclose copies of each of the following documents:

1. Two copies of the Specification document which incorporates Preliminaries, Specification and Schedule of Works
2. Two copies of each of the drawings as detailed on the attached drawing issue sheet.
3. Ditto with consultants' drawings*
4. Bills of Quantities *
5. The Employer's Requirements *
6. Pre-construction Health and Safety information * (if not issued directly by the CDM Co-ordinator)
7. One further copy of the Tender Summary and Form of Tender
8. An address label for the return of the Tender.

It is intended that the works should start on site as soon as possible after the acceptance of the tender, and your Form of Tender should indicate the earliest start date.

If you wish to visit the site, please make arrangements for inspection during the tender period by telephoning the Employer on the number listed in the Preliminaries.

Further drawings and details may be inspected at .... *

The completed Form of Tender with Tender Summary and priced Specification document (and/or Contractor's Proposals specified in the tender documentation*) are to be sealed in a plain envelope with the address label provided and delivered or sent by post to reach this office no later than 12:00 noon on XXXXXXXXXX 20... There should be no name or markings on the documentation which indicates the sender. (Or alternative for e- tendering.*) Tenders received after this time will not be considered and will be returned unopened.

The tendering procedure will be carried out in accordance with the JCT Tendering Practice Note 2012 using Alternative [insert 1 or 2] for

(continued)

Figure 4.7 Example invitation to tender letter

Tender invitation letter (*continued*)

correcting errors. The Employer reserves the right to postpone the date for the return of tenders and it is to be understood that neither the lowest nor any tender will necessarily be accepted. No tender expenses will be payable.

Tenders will be assessed on the basis of lowest price/best value* criteria as set out in the preliminary invitation to tender information attached. You are reminded of the need for confidentiality and your agreement not to divulge your tender price to others.

Will you please acknowledge receipt of this letter and confirm that you are able to submit a bona fide tender in accordance with these instructions. Should you have any queries during the tender period please contact xxxxxxxxxxxxx or xxxxxxxxxxxxx who will be dealing with this project.

Yours faithfully

Confirmation of receipt of documents

It is important that you obtain an acknowledgement from each tenderer, particularly when the documents have been issued by email. In such instances, resending the email letter without the attachments with a request for confirmation of receipt usually suffices.

Concurrent with the tender issue, the client should be sent a copy of all the documents with an accompanying letter confirming that the tenders have been issued and advising on the need for confidentiality. An example letter is shown in figure 4.8.

The contractors should notify the contract administrator of any deficiencies in the information issued, particularly in the case of design and build tenders. During the tender period any queries from contractors should be dealt with promptly. Each query and the subsequent response should be circulated to all the tenderers to ensure that all tenders are priced on the same basis.

Tender issue letter to client

Dear Sirs

Re: PROJECT TITLE

I am pleased to advise that tenders have now been issued and I enclose a copy of the documentation for your information. Should you have any queries on these we can issue further tender information during the tender period.

There are XX tenderers and tenders are due back by 12:00 noon on XXXXXXXX 20.. which is in 3 weeks' time. This is less time than I suspect the contractors will want because they sometimes have difficulty getting prices from their mechanical and electrical subcontractors, therefore it is possible that we may need to extend this deadline if sufficient tenderers ask. The tenderers should contact you to arrange a site visit and under the terms of the selective tendering rules you should not disclose the names of any of the tenderers to them, should they ask, to ensure confidentiality.

Please do not hesitate to contact me if you have any queries or need any further clarification and I will be in touch as soon as tenders are received, unless I hear from you in the interim.

Yours sincerely

Figure 4.8 Example tender issue letter to client

If it is necessary to issue any further information during the tender period then this should be sent out to all tenderers as soon as possible to ensure that they have time to include it. Late information could lead to requests for an extension of the tender period, which would be difficult to refuse in such circumstances.

Any information issued during the tender period should be accompanied by a request that the contractors acknowledge receipt and confirm that the further information will be included in their tender prices.

Tender return

All tenders should be returned by the stated date and time. It is prudent to give a tenderer who has not replied a reminder a few hours before the deadline to ensure that their tender will be in on time. The use of coded return labels, as noted above, is helpful for this. Late tenders should not be accepted and should be returned unopened. Tenders should also be rejected for non-compliance with tender instructions.

The delivery address for tenders should have been agreed from the outset. Usually they will be sent to the contract administrator, but if they are to go to the client, it is imperative that the client, and more importantly the client's staff, are fully aware of the correct procedure and the need for confidentiality. When delivered by hand, the contractor should be given a receipt, which should be dated *and* timed, and the tender document itself should have the time written on the envelope *which should not be opened until the formal tender opening.*

Contractors will often submit their tenders electronically by email or fax before the due time, which, in part, defeats the confidentiality issue. However, if this is accepted then the tender should not be seen by the contract administrator but printed and put in an envelope and marked as a 'tender for…..' and placed with the other tender returns for opening at the appropriate time.

Each tenderer should be notified immediately following the closing date for the receipt of tenders, or no later than 15 days after receipt, that their tender has been received and whether it is under consideration or not. In addition they should be advised that they will be notified of the names of the other tenderers and the prices received once the tenders have been evaluated and the contractor has been selected.

Tender opening and assessment

Tenders should be opened as soon as possible after the tender due time and be checked for any missing information, such as priced schedules, method statements and so on.

Tenders should be opened in the presence of at least two persons. Often the client will want to be there, but in many instances the client will leave it to the contract administrator to open the tenders and report subsequently. However, if the client is to open the tenders, the contract administrator should preferably be there to ensure that the procedures are respected and to give their initial comments.

The client should be advised that while they are not obliged to accept the lowest tender, a decision not to accept the lowest *without good reason* could upset the other contractors and could prevent them tendering on future projects.

The person appointed to open the tenders should open each one in turn and, in the first instance, write down the prices, start dates and contract periods so that an initial evaluation can take place. An example of a tender received schedule is shown in figure 4.9, which should be signed by all present as a record of the tenders submitted. If the client is not present then the contract administrator should contact them by phone to advise of the range of tenders and timescales pending evaluation and the issuing of a formal tender report. In the meantime, the tenders received schedule can be forwarded to the client.

The tenders should then be looked at in more detail.

If a contractor has qualified their tender, the tender will be non-compliant. The contractor should be asked to withdraw the qualification with no alteration in price or withdraw the tender. If they refuse, the tender may have to be rejected.

The priced schedules for the two lowest tenders should be requested, if they are not already part of the tender return documentation. They should then be checked arithmetically and to ensure that the contractor has priced for every element.

A check should also be made for any provisional sums, which may have been included if the tenderer had not left sufficient time for pricing or did not get subcontract prices in on time. Tenders that include contractor-generated provisional sums should not be accepted, but if alternative 2 (see below) is used, the contractor could be given the option of firming up the price.

Any errors identified during the arithmetical and technical checks of the schedules should be noted. The contract administrator should then ask the tenderer to address the error(s). The correction procedure will depend on what provisions were included in the tender documents. Under the JCT's *Tendering Practice Note 2012* there are two alternative scenarios:

- *Alternative 1* is that price prevails: the tenderer should accept the error or withdraw.
- *Alternative 2* is that the tenderer can accept the error or amend any genuine errors – there is a risk to the contractor that their tender will no longer be the lowest or best value and will thus be discounted.

Schedule of tenders received

File No: 5210

SCHEDULE OF TENDERS RECEIVED

CLIENT: Mr and Mrs A Smith

PROJECT: Extensions and alterations at 41 York Way, Sometown, ST4 4TS

Name of Contractor	Start date/timescale	Amount of Tender
1) Building Partners	7 January + 26 weeks	£258,246.00
2) Gray Builders	7 January + 26 weeks	£257,789.50
3) Builder Bros	7 January + 26 weeks	£250,000.00
4) John Briggs	27 January + 28 weeks	£254,183.00
5) Bricks and Mortar	7 January + 26 weeks	Tender invalid
6)		£

TENDERS OPENED BY: ....

Client

....

Contract Administrator

....

DATE:

27 November 2012

Figure 4.9 Example of a schedule of tenders received

Adjusting tender prices

If a tenderer has accepted an error under alternative 1 but their tender is still the lowest and is accepted, the priced documents should be endorsed such that all rates and prices (excluding preliminaries, contingencies, provisional sums and the like) should be increased or decreased in proportion to that figure. This should be signed by both parties.

The same will apply under alternative 2 if the tenderer chooses to stand by their figure rather than correct a genuine error.

The contract administrator should also:

- check that all items have been allowed for and included in the tender submission
- evaluate tenders against the selected criteria
- check any method statements
- review the contract programme
- arrange for the health and safety adviser to check the contractor's construction phase health and safety plan
- on design and build tenders, review the design drawings, specification and Contractor's Proposals
- appraise specialist tenders.

At this stage the client and the contract administrator may wish to interview the preferred tenderer(s). In the case of traditional contracts this should be unnecessary as the tenderers would have gone through the pre-selection process and so the contract administrator will already be satisfied that all the contractors are capable of undertaking the works. However, in the case of design and build or management types of contract, interviews are necessary to allow the tenderers to explain their proposals in more detail, clarify any points in their submissions and introduce key personnel.

The contract administrator should be wary of very low tenders – these should be checked even more thoroughly for errors and missing items. If the tenderer is satisfied that the figure is correct, the client should be warned of the risks of accepting such a low tender. If there are no apparent errors then it is possible that the contractor is trying to buy the work to keep their workforce going during a difficult time. This can lead to the contractor cutting corners with the design/construction, or they could go into liquidation during the contract period. This will impact on the client through additional costs for retendering, extra consultants' fees and the inevitable costs of the new contractor gearing up and putting right any defective or incomplete work.

Tender report

A report on the tenders received should be prepared which includes a recommendation on which tender to accept. This can range from a full report by the cost consultant to a brief report in the form of a simple letter, such as in figure 4.10. The tender report/letter should be sent to the client together with copies of the tenders received and a request for written instructions to proceed.

Where the overall budget is the overriding factor, it is sometimes more appropriate when reporting tenders to include an overall project cost as part of the tender report. The costs of consultants, statutory fees, reports, etc., as well as VAT (where the client is unable to claim it back) should be added to the tenders.

What if the tender is over budget?

If the preferred tender figure is above the client's budget, the contract administrator has a number of options for the client to consider:

- increase the budget by using reserves or borrowing more
- reassess the scheme to see what can be omitted without affecting the overall requirement
- negotiate or retender any cost reductions
- abandon the tender and retender with another scope of works.

It is common that a client who wishes to pursue cost reductions will negotiate only with only the lowest tenderer. However, given that all tenderers should be treated fairly and equally, technically all tenderers should be given the opportunity to price the reductions. Indeed, this is essential on public sector contracts. In practice much will depend on the level of cost reduction sought, but, particularly if tenders are close, it is prudent to obtain new costs from the lowest two or three contractors.

All changes to the drawings, specification, tender documents and pricing should be fully documented and recorded.

Tender report letter to the client

Dear Sirs

Re: Extensions and alterations at 41 York Way, Sometown, ST4 4TS

Further to the receipt of tenders and our subsequent telephone conversation I can now confirm the details of the tenders received.

All five tenders were received by the appointed time, but one was deemed invalid, as follows.

– Bricks and Mortar	tender invalid
– Building Partners	£258,246.00
– Gray Builders	£257,789.50
– John Briggs	£254,183.00
– Builder Bros	£250,000.00

These figures exclude Value Added Tax at the Standard rate but include provisional sums and contingencies which may or may not be expended.

We have checked the priced schedules of the two lowest tenders submitted by John Briggs and Builder Bros and find them to be arithmetically correct but we have made a number of observations.

John Briggs had made an ...

Builder Bros have allowed for ...

Builder Bros have stated that they would be able to undertake the work starting on 7 January 2013 with a 26-week contract in accordance with the timescale set in the preliminaries whereas John Briggs can only start on 27 January 2013 and will need 28 weeks to complete the works.

ADD ANY OTHER JOB-SPECIFIC ITEMS

Whilst you will appreciate that all the tenders are over the £225,000.00 budget I would advise that these figures include a £15,000.00 allowance for a GSHP installation which was added at a late stage as well as a contingency sum of £10,000.00 against any unforeseen items which may or may not be expended. This consideration makes the tenders more comparable with the budget.

(continued)

Figure 4.10 Example of a tender report letter to the client

Tender report letter to the client (*continued*)

I enclose a copy of the priced schedules of work for the two lowest tenderers together with copies of the other tender forms for your information and I look forward to receiving your instructions at your earliest convenience.

Following receipt of a written instruction to proceed, the next step will be to arrange a Pre-Contract meeting to meet with the successful contractor with a view to starting on site as soon as possible after that date.

Please do not hesitate to contact me if you have any queries or you need any further information.

Yours sincerely

New approaches to contractors

If the tender prices received are too high, clients will sometimes ask for further tenders to be invited from smaller or different contractors, with the sole aim of getting a better price. While it is difficult for the contract administrator to refuse such a request, it should be resisted if at all possible.

In one instance, a committee which asked for this was advised by the contract administrator that it was not in accordance with professional procedures and that, while he had no doubt the committee had the right intentions, he would be unable to convince the original contractors of his own good faith for future projects. A subtle but firm response!

Tender acceptance and award

The tenders will need to be accepted within the period stated in the tender documents or they will 'lapse'. If no period is stated in the tender documents they will lapse after a 'reasonable period'. This is open to interpretation and therefore stating a figure such as 6 or 13 weeks is a much better option. Once the tenders have lapsed then the client can either retender the project or negotiate with the lowest tenderer(s).

Once the client has decided on which contractor they will use to undertake the works, the successful contractor should be advised immediately of the intention to place the contract with them. The most practical way is to telephone them followed up with a letter – commonly known as a *letter of intent* (see figure 4.11). At the same time dates for the pre-contract meeting can be discussed and, if not already available, the contractor can be asked to prepare their construction phase health and safety plan. It is also good practice to advise the selected contractor of the names of the other tenderers and the prices received.

On public sector contracts, before the Building Contract can be formally awarded there is a 'standstill period', which gives the other tenderers the opportunity to question, be debriefed or challenge the decision if they have any concerns about the decision-making process. Under the Public Contracts (Amendment) Regulations 2009, a Contract Award Notice should be posted in the *OJEU* within 48 days of the award.

What is a letter of intent?

A letter of intent is an expression of an intention to enter into a contract at a future date without creating a contractual relationship until the contract has been entered into. It is not an 'agreement to agree'. The term does not have a legal meaning, such as *subject to contract.*

The intention to enter into a contract does not give rise to any legal obligation or any liability in contract law, but it does not exclude or negate a right to recover reasonable expenditure on a quantum meruit basis.

Letters of intent are sent at a time when it is anticipated that the contractor will start incurring costs and overheads as an interim arrangement to mobilise the contractor prior to a formal contract being signed but they should not be seen as an alternative to the signing of the formal contract.

That said, the advice to clients should be that it is preferable to avoid sending letters of intent because they do not cover all the eventualities set out in a standard contract. They also reduce the pressure on a contractor to sign a more comprehensive set of obligations and allow them to renegotiate if any unforeseen risks manifest themselves during the early days of the contract, before the formal contract is signed.

Letter to successful contractor

Dear Sirs

Re: Extensions and alterations at 41 York Way, Sometown, ST4 4TS

Further to our telephone conversation on XXXday I am able to advise you that you were the lowest tenderer and that we have been advised by Mr and Mrs Smith that they wish to accept your tender in the sum of £250,000.00 plus VAT. On this basis I confirm that we have arranged a Pre-Contract meeting for XXXXX 20.. at 1:00 pm at the property and I enclose an agenda for your information.

I look forward to seeing you then but in the interim please treat this letter as a formal Letter of Intent to place the Contract with you and (insert on CDM projects) I would be grateful if you could prepare your Construction Phase Health and Safety Plan so that this can be approved before the start date. The works will start on Monday xxxxxx 20.. with a xx week contract period although you indicated that you may be able to start earlier, on the xxxxx, but this can be discussed at the meeting.

We will issue the contract documents for signature shortly so they can be completed before the date for commencement. In the meantime, for your information, I would advise you of the tenders received in descending order and the tenderers in alphabetical order.

One tender invalid
- £258,246.00
- £257,789.50
- £254,183.00
- £250,000.00

- Bricks and Mortar
- John Briggs
- Builder Bros
- Building Partners
- Gray Builders

I look forward to hearing from you and to working with you on this project.

Please do not hesitate to contact me if you have any queries or you need any further information.

Yours sincerely

Figure 4.11 Example of a letter to the successful contractor

Advice to unsuccessful contractors

At the same time as issuing the letter of intent to the successful contractor, the contract administrator should notify the other tenderers that they have been unsuccessful (figure 4.12). Good practice is to advise them of the names of the other tenderers and the prices received.

Example letter to unsuccessful contractors

Dear Sirs

Re: Extensions and alterations at 41 York Way, Sometown, ST4 4TS

Further to receipt of your tender submission for the above works, we thank you for this but in this instance have to advise that it was not successful.

At the time of tender return 4 tenders were received and I would advise you of the tenders received in descending order and the tenderers in alphabetical order

- One tender invalid
- £258,246.00
- £257,789.50
- £254,183.00
- £250,000.00

- Bricks and Mortar
- John Briggs
- Builder Bros
- Building Partners
- Gray Builders

Following a discussion with the client it was agreed that the tender received by Builder Bros would be accepted and the Contract is to be awarded to them.

We thank you for your submission and hope that you will be happy to be considered for future tenders when the outcome may be different.

Please do not hesitate to contact me if you have any queries or you need any further information.

Yours sincerely

Figure 4.12 Example of a letter to unsuccessful contractors

The contract administrator should report the tender returns in such a way that it is not possible to see what actual figure a particular tenderer submitted (although it is not uncommon to advise who the successful contractor was). This is most commonly done by listing the figures in ascending (or descending) order, with the names of the tenderers listed alphabetically.

Preparation of the contract documents

As far as practical, the Building Contract should always be completed and signed before the works starts on site. Both the client and the contractor should sign and witness the contract. Often the time taken to collate the necessary paperwork can be outstripped by the desire to start construction. If this happens it can become harder to resolve any disputes on the contract details. Indeed, there have been cases where the courts have had to rule on an implied contract as the Building Contract has remained unsigned.

On a traditional, fully-designed project, the contract documents may include:

- the articles of agreement and conditions of contract
- contract drawings, which are usually the tender drawings
- bills of quantities
- specifications
- schedules of works
- a schedule of tender adjustments or clarifications negotiated and agreed after the receipt of tenders and prior to the signing of the contract.

On design and build projects, the contract documents may comprise:

- the articles of agreement and conditions of contract
- the Employer's Requirements
- the Contractor's Proposals
- the contract sum analysis
- bills of quantities for elements of the design.

The contract documents should be assembled using the priced copies of the schedules, bills of quantities, etc. Any subsequent tender issue(s) and/or queries should be copied and added to the front of the specification to ensure that there is a full record of all the information upon which the tender sum was based. In addition, if the contract sum is not the same as the tender sum then there should be a list of items adjusted and priced to show how the contract sum was calculated.

The articles of agreement and conditions of contract

These items should be completed using the standard form appropriate to the type of contract being used and noted in the preliminaries. The contract administrator has the choice to complete the contract forms in electronic or hard copy formats. The details and terms of how the form should be completed will have been set out in the contract preliminaries issued with the tender.

Who is party to the Building Contract?

The contracting parties to the contract are the client (now known as 'the Employer' under the terms of the Building Contract, although the RIBA Plan of Work 2013 continues to refer to the 'client') and the main contractor. When completing the contract it is extremely important that the correct names, addresses and company numbers, where appropriate, are used on the Building Contract documentation to ensure that the named companies or individuals are *actually* the contracting parties and have the authority to sign the documents. However, if either of the parties is using another address, for example their trading address is different to their company address, then this should be noted within the contract documentation.

This can be a common problem with building contracts where they have been concluded in the name of a person or trading name without proper consideration having been given to who the contracting party is.

This situation can equally apply in the case of professional services contracts. The problem is that if the client considers there is something wrong with the level of service you provided, they can refuse to pay your fee because your fee agreement bears the wrong company name. Indeed, it is just as important, if not more, that you know the person or company responsible for paying the accounts.

Who is party to the Building Contract? (*continued*)

The person who signs the document will generally be the contracting party, unless they clearly do so as an agent for a suitably identified company. Therefore, you should check and note not just the identity of the contracting parties, but also the capacity in which the signatories will sign, particularly in a corporate situation. It is also important, particularly with group structures (where company names can be similar) and generic trading names, which may confuse the position further, that the corporate structure is understood and, if necessary, clarify this in writing with them. For the avoidance of doubt, you should also ensure that the relevant company number(s) is included as well as the correct name within the document.

The contract form should be completed *exactly* as is stated in the preliminaries and every amendment or insertion should be marked with a pencil cross so that they can be initialled by both parties as confirmation that they are acceptable. In practice it is not uncommon for some to be missed, so it is useful to highlight these with small sticky notes labelled 'Initial' and a larger one on the signatures page labelled 'Sign and witness'.

The drawings

All tender drawings should be copied and each one should be referenced for the parties to sign. Each drawing should be identifiable as part of the overall package and marked '1 of X', '2 of X' and so on. Practices may have standard rubber stamps for paper documents and CAD blocks for digital documents.

The specification, schedule of works and other documents

These items should be assembled with the prices included, together with the contract sum breakdown and all the tender variations as described above. The documents should be referenced and signed in the same way as the drawings.

Two sets of the documents should be assembled, with one being the original for signing and the other being the *certified true copy*. However, on smaller projects, both parties can sign both sets so that there are effectively two originals.

Arranging for signature

Following the assembly of the sets of contract documents, the contract administrator should arrange for their signature by both parties, either by hand or as a deed (figure 4.13). The simplest and most expedient way of arranging this is for both parties to meet at a convenient place and time. The documents can then be signed and witnessed, and also checked for errors at the same time by the contract administrator before they are dated.

If a meeting is not a requirement or not convenient, the contract administrator should send, by post, courier or hand delivery (the delivery

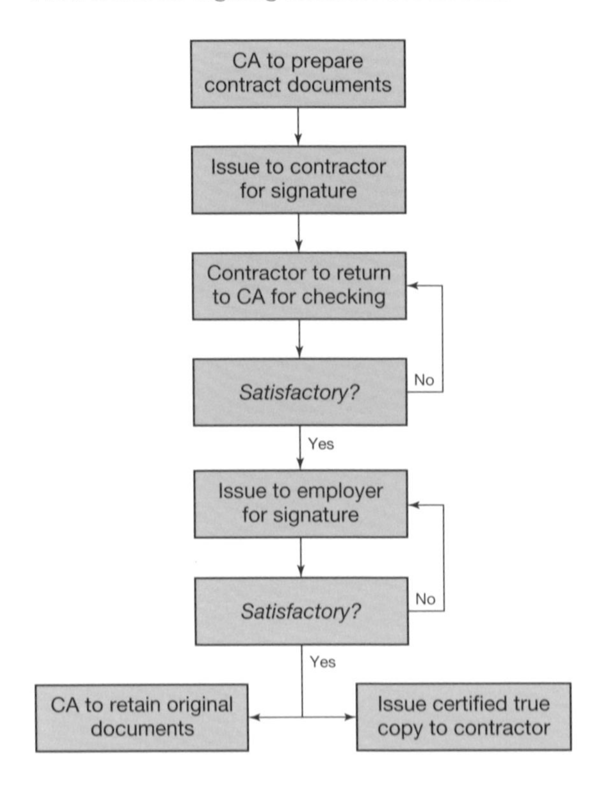

Figure 4.13 Flow chart for signing contract documents

method should be trackable), the documents first to the contractor, with an accompanying letter asking them to sign and witness the documents (and indicating how this should be done) but stating that they should not date the documents. An example of a suitable letter is shown in figure 4.14.

On receipt of the signed documents from the contractor, the contract administrator should carefully check that the documents have been

Letter to contractor issuing contract documents

Dear Sirs

Re: PROJECT TITLE

Further to our pre-contract meeting please find enclosed the original copy of the Contract Documents. They comprise the following:

1. JCT Minor Works Building Contract MWD11. This needs to be signed on page 13 by either a Director or Company Secretary. Please DO NOT date the Contract.

   Please also initial all entries and amendments where indicated with an 'X'.

2. The Contract Specification also requires the signature of a Director on the front cover. Again please DO NOT date.

3. Architect's drawings 5210/01-04 inclusive each requiring the signature of a Director.

   Please DO NOT date.

4. [Add any further documents.]

Once the documents have been signed can you please return them to this office for forwarding on to the Employer.

Following signature by the Employer you will receive a certified true copy of the documents and the originals will be retained in this office for safekeeping and record.

In the meantime, should you have any queries then please do not hesitate to contact me.

Yours sincerely

Figure 4.14 Letter to contractor issuing contract documents

signed, witnessed and initialled in the appropriate places. Once the contract administrator is satisfied that the contractor has completed the documentation correctly, the documents should be sent to the client by the same means with a similar, amended letter.

The client should return the signed but undated documents to the contract administrator for checking. Once the contract administrator is satisfied that the documents have been completed correctly by both parties, the agreement and the other documents should be dated.

Unless the client wishes to keep the contract documents (in which case the contract administrator should take a copy), the original documents are usually retained by the contract administrator for safekeeping and this should be confirmed to both parties. The documents should be kept in a secure and fireproof location.

A copy of the documents should be sent to the contractor for their records. Unless two sets of the documents have been signed, the spare copy of the documentation will be engrossed by the contract administrator as a certified copy. As before, the documents should be endorsed stating that the document is a 'Certified True Copy' and be signed and dated by the contract administrator.

A certified true copy of the signed agreement will be also required. A completed, but unsigned, copy can be used, but a simpler way is to photocopy the signed agreement and bind it securely so that each party effectively has a signed version. Whichever way is preferred, the contract administrator should endorse the document with the wording 'We certify this to be a true copy of the Within Written Agreement dated ... etc.'

The contract administrator should then sign and date all the documents and send them to the contractor with a suitable covering letter. At the same time the contract administrator should copy this to the client to advise the client that this has been done.

What about change control?

Change control on a project should be fully documented irrespective of the project stage as any changes to a project can have an impact on time, cost and quality. The later in the development of the project the changes occur, the greater that impact is likely to be.

Design change notice

Ed Associates Ltd.
141 New Business Park, Anytown, AT51 1ST

Job no: *EA123* Job title: *NEW MUSIC CENTRE*

DESIGN CHANGE NOTICE AND RECORD

To:

The Governors of AGood School
London Road
Anytown
AT1 1TA

Enclosures:

Please find enclosed the documents listed below.
Please enter comments, photocopy, and return original notice by

Date 20dec12

Confirming your request for additional network points and electric sockets as shown on drawing no. 5210/006B attached.

Issued by: Ed **Date:** *20dec12*

Comments

I agree to this amendment

James Scott
Head Teacher

Signed: **Date:** *10jan13*

Implementation record

Change adopted: YES/ ~~NO~~ **Included in contract**: YES/ ~~NO~~

Covered by Architect's Instruction **No.** *6* **Issued date:** *21jan13*

Figure 4.15 Example of a design change notice

During the pre-construction stages design documents will be issued to the client for approval. Once the client has given approval, a Change Control Procedure should be introduced for any further variation to ensure that the amendments have the express permission of the client. On design and build and management forms of contract the 'client' could equally be the contractor.

At Stage 4 Change Control Procedures will be required for all members of the design team when working up the detailed design and by the contract administrator when the tender documentation has been prepared as indicated in the example in figure 4.15, but the form shown can be used at any stage.

Chapter summary 4

This chapter has described the tasks and procedures that the contract administrator completes leading up to the award of the Building Contract. It has primarily considered the traditional procurement route, but has also covered minor variations depending on the selected form of contract. Some projects may require tendering to be conducted at an earlier stage, but regardless of the stage at which tendering is undertaken, the same rules will apply.

Although other members of the team will be performing roles during this period, the contract administrator has the key role in ensuring that fair and competitive tenders are obtained, to enable the client to obtain the lowest price or best value from the selected contractors.

Throughout this process the contract administrator needs to be methodical, accurate and thorough.

Irrespective of the stage at which the project is tendered, the tendering process and the award of the Building Contract should be completed by the end of the appropriate stage for the form of procurement so that construction works can be started in earnest at Stage 5.

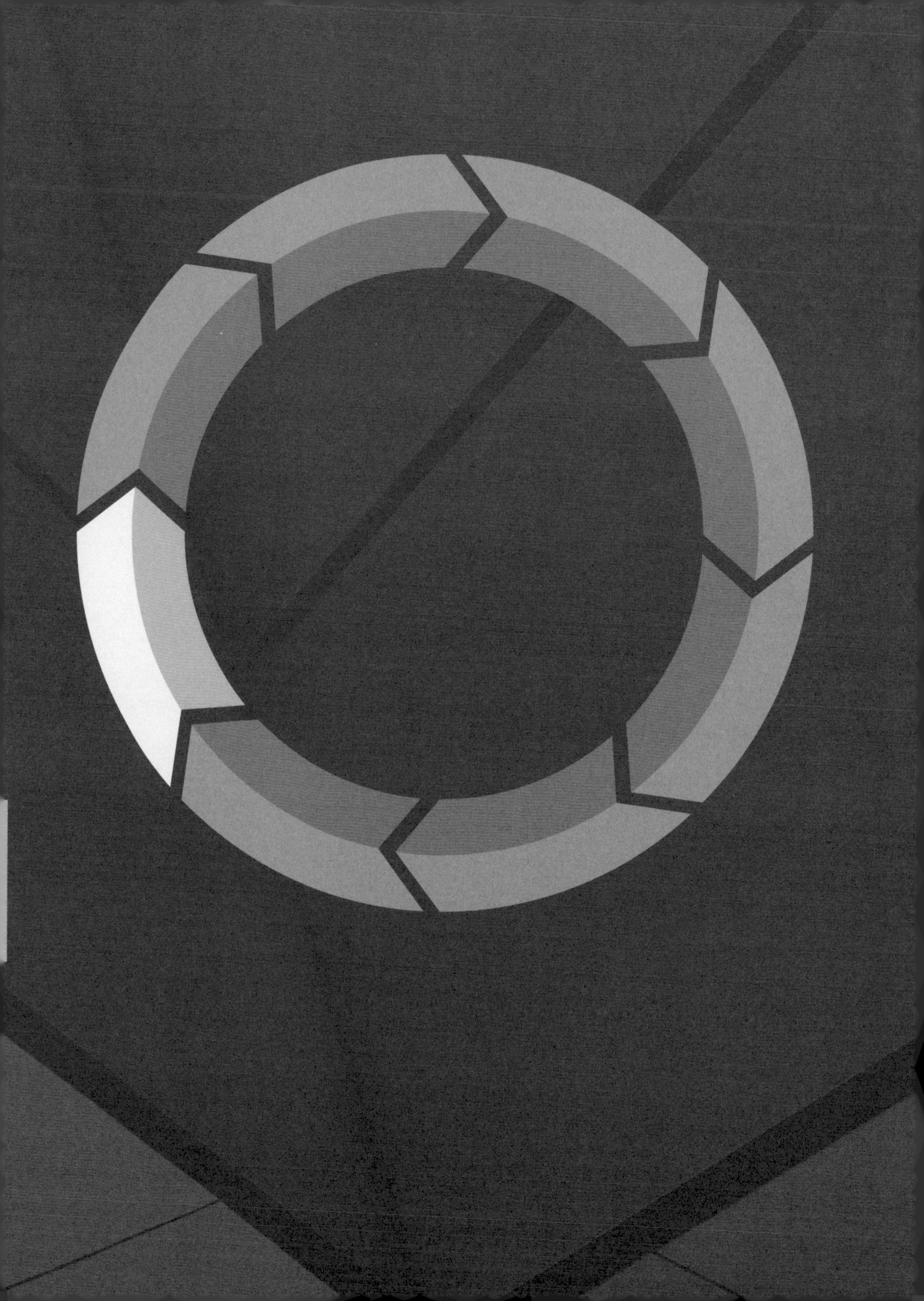

Stage 5

Construction

Chapter overview

During Stage 5 the building is constructed on site in accordance with the Construction Programme and this is where the majority of the contract administrator's role is performed.

This chapter considers the duties and obligations of the contract administrator, as well as the roles and procedures which may arise during this stage together with the best methodologies for dealing with them to ensure that the building is completed in accordance with the Building Contract.

Irrespective of the contract administrator role, the Schedule of Services for the design team members will have set out the designers' duties to respond to Design Queries from site, to carry out site inspections and to produce quality reports. However, on a small project it is likely that the contract administrator will be undertaking a number of these project roles.

The key coverage in this chapter is as follows:

What is the role of the contract administrator at Stage 5?

What are the tasks and procedures before work starts on site?

What is the Construction Programme?

How is the contractor instructed?

What are the tasks and procedures after work start on site?

What are the tasks and procedures for payments?

What if the works are delayed?

What happens at Practical Completion?

Introduction

The main role of the contract administrator during Stage 5 will be to facilitate the successful monitoring of the construction process by carrying out activities between the signing of the Building Contract and Practical Completion at the end of the stage. When considering the actions needed at Stage 5 the contract administrator should be aware that there is an overlap with Stage 6, as can be seen in figure 5.1.

Therefore, during the period leading up to handover it should be noted that, because of this overlap, sections of the chapter covering Stage 6 will also be relevant during the latter part of this stage.

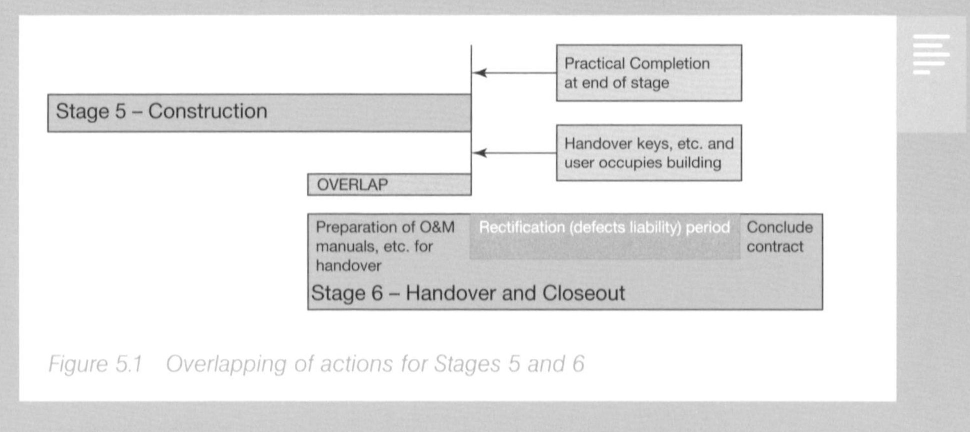

Figure 5.1 Overlapping of actions for Stages 5 and 6

From figure 5.1 it can be seen that although Practical Completion is achieved at the end of Stage 5, the actions required to facilitate the Handover Strategy undertaken during Stage 6 starts earlier, with the assembly of information and documentation needed to certify Practical Completion. This is discussed in more detail on page 194: Stage 6.

Practical Completion will be achieved either as a single event or in a phased manner, whereby Practical Completion will be reached in stages (with contracts using the sectional completion option). The tasks and procedures in this chapter are based on a single completion date, but, should the terms of the contract require sectional completion, the same tasks and procedures should be undertaken for *each* section.

Unlike earlier stages, where the procurement route will impact on the tasks to be performed by the contract administrator, in Stage 5 the tasks will generally be common to all forms of procurement in respect of the contract administrator role. On design and build projects the contract administrator role will be undertaken by the employer's agent.

The procedures generally relate to JCT forms of contract, but there are some minor variations with other forms. These have been highlighted within the chapter.

What are the Core Objectives of this stage?

The Core Objectives of the RIBA Plan of Work 2013 at Stage 5 are:

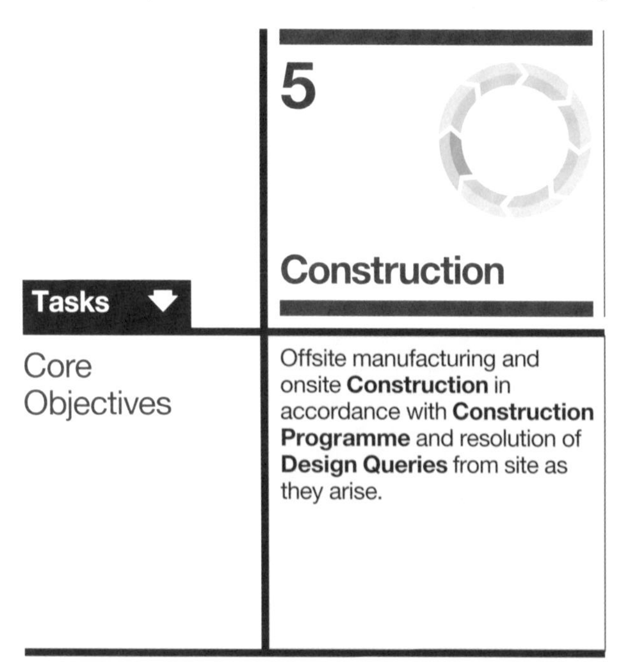

The Core Objectives at Stage 5 revolve around the offsite manufacture and onsite construction in accordance with the Construction Programme and the resolution of Design Queries from site as and when they arise.

What supporting tasks should be undertaken during Stage 5?

The Suggested Key Support Tasks noted in the RIBA Plan of Work 2013 have been devised to support the Core Objectives and to ensure that the procedures and documentation required to carry out the administration of the Building Contract are in place so that the building is completed to enable the client to take it over. At this stage the Suggested Key Support Tasks for the contract administrator are:

- Review and implement the Handover Strategy, including agreement of information required for commissioning, training, handover, asset management, future monitoring and maintenance and ongoing compilation of 'As-constructed' Information.
- Update Construction Strategy.

What is the role of the contract administrator at Stage 5?

The tasks required of the contract administrator vary, albeit only slightly, depending on the form of procurement. However, the contract administrator's relationships with other parties may vary according to the procurement route.

On traditional contracts the contract administrator should always act in a fair and reasonable manner and must be impartial in dealings with the client and the contractor. In contrast, on design and build projects, the contract administrator (the employer's agent) is not impartial as their authority comes from the client (as the employer), not the contract. Furthermore, on the different forms of management contract the contract administrator is again an independent consultant, but the contractual obligations differ from the traditional form and the contract administrator's roles and responsibilities should be fully described in the Building Contract.

In traditional forms of contract the contract administrator is appointed as a third party consultant to act on the client's behalf, undertaking a number of administrative functions which will include the following

Stage 5 tasks:

- administering the contract and Change Control Procedures
- checking specialist drawings for contractor's designed portion
- seeking instructions from the client in relation to the contract
- issuing instructions, such as variations or relating to prime cost sums or defects
- chairing construction progress meetings
- preparing and issuing construction progress reports
- coordinating and implementing site inspections
- considering claims and extensions of time
- evaluating payments and issuing interim certificates
- agreeing commissioning and testing procedures
- agreeing defects reporting procedures
- ensuring that project documentation is issued to the client
- issuing certificate(s) of Practical Completion.

Once the Building Contract is entered into both parties will have rights, duties and obligations. The contract administrator should check the relevant standard form of contract, which will set out the procedures and processes that must be observed.

The contract administrator should have a robust and efficient system for monitoring the project and for logging and following up outstanding information. This will also provide evidence to show that the project has been managed properly – according to the RIBA Insurance Agency, one-third of all professional indemnity insurance (PII) claims relate to contract administration.

The time limits for the issue of certificates, notifications, etc. stipulated within the contract are crucial and the contract administrator should adhere strictly to these to avoid the contract becoming 'at large'. If that happens, the client is unable to claim liquidated and ascertained damages from the contractor, which could lead to a claim against the contract administrator.

Time at large

The vast majority of construction contracts include a date by which the works described in the contract should be completed: the date for Practical Completion.

The term 'time at large' describes a situation where no date has been set for completion, or where the date for completion has become invalid, thus the contractor is no longer bound by the obligation to complete the works by the date set.

If there is no clear completion date set or if the contract does not allow the construction period to be extended then time is at large from the outset. Alternatively, where a completion date was set by the contract, time can also cease to apply either by agreement or by an intervention, such as an instruction for additional work that would take the project over the agreed timescale and for which the client has not allowed additional time.

The contract administrator (on traditional contracts) has a function independent of both the client and the contractor. That said, the contract administrator effectively has a dual role as they are also the employer's agent.

This dual role can sometimes give rise to conflicts. A typical example would be where the client, wanting to limit their cash flow, looks to the contract administrator to under-certify or not agree to a fair and reasonable price for additional work when putting together the final account. Such interference by the client, preventing the contract administrator from acting fairly in contractual matters, will cause the client to be in breach of contract. It is therefore of great importance that the client is fully aware of this and that the contract administrator manages the position openly and correctly.

However, this impartiality has a major benefit for both the client and the contractor as either party may challenge any decision made by the contract administrator.

Where do the contract administrator's obligations lie?

Often, the client sees the contract administrator as their tool for punishing the contractor, whereas the contractor largely understands the role of the contract administrator and has much greater knowledge of the contract administrator's obligations under the terms of the Building Contract.

In reality it is a difficult balance, particularly as the contract administrator's appointment rests with the client. However, the client must be made fully aware that in respect of administering the terms of the contract, the contract administrator must sit 'squarely in the middle'.

What is the role of the employer's agent on a design and build project?

In construction the term 'employer's agent' is used to describe an agent acting, on behalf of the employer, as the contract administrator for design and build contracts. The employer's agent is likely to be either the project lead or the cost consultant; however, the role can be carried out by someone from the client organisation, such as an in-house project manager, or by an independent project manager appointed by the client.

After the contract has been awarded, the employer's agent's role as contract administrator during Stage 5 includes:

- issuing instructions
- coordinating the review of information prepared by the contractor
- considering items submitted by the contractor for approval, as required by the Employer's Requirements
- managing Change Control Procedures
- reviewing the progress of the works and preparing reports for the client
- validating or certifying payments
- considering claims
- monitoring commissioning and inspections
- arranging handover
- certifying Practical Completion.

What administration system could you use?

The contract administrator should have a robust and efficient system for administering the Building Contract based on the particular form of contract selected. The examples shown later in this chapter are based on the RIBA Contract Administration Forms for use with the JCT Standard, Intermediate, Minor Works and Design and Build contracts. They are issued in hard copy or electronic formats using *NBS Contract Administrator*. Further guidance and detailed information can be found by contacting the NBS:

Telephone: 0845 456 9594 (option 2)
Email: sales@theNBS.com

Contact details for other contract forms are available from the contracts' publishers (see page 46).

What are the tasks and procedures before work starts on site?

Following the successful tendering and appointment of the contractor at Stage 4 (or earlier) and before the work starts in earnest on site, the contractor will be mobilising and organising the resources required or organising any offsite manufacturing. The contract administrator should allow the contractor a reasonable time to prepare as it can be counterproductive if the contractor starts on site too soon after being appointed.

At the commencement of the stage the contract administrator should advise the client, as the employer under the contract, of their role and their responsibilities for the works on site, which may vary according to the type of procurement. These should include:

- ensuring that the site will be available on the date set for the commencement of the contract works – if it is not, there may be repercussions for the contractor, leading to an immediate delay to the Construction Programme

- ensuring that the construction phase plan required under the CDM regulations is sufficiently advanced, otherwise work cannot start on site
- ensuring that works are insured
- knowing the level of involvement they should have with the contractor on site and the implications of giving instructions other than through the contract administrator
- ensuring that payments are made to the contractor in accordance with the timescales set, following the issue of certificates by the contract administrator
- organising and paying for materials, products and services which are outside the scope of the main contract, such as specialist installations, fittings, etc.
- understanding procedures for accepting handover at Practical Completion.

Checking insurances

Insurance requirements for the project will have been discussed with the client prior to tendering. Before construction commences the contract administrator must ensure that the necessary insurances are in place.

The wording for the insurance requirements should be checked in the particular form of contract being used, but it will generally fall into one of three categories:

- insurance by the contractor in joint names
- insurance of existing structures by the employer in joint names
- insurance of existing structures in own name.

The contract administrator should obtain evidence that the above insurances are in force for the duration of the project, together with the details of the contractor's public liability insurances. The dates for renewal should be noted – if they expire before the contract completion date then the party responsible for the insurance should be asked to provide renewal certificates at the appropriate time.

Arranging the pre-contract meeting

As soon as possible after the contractor has been appointed the contract administrator should arrange a pre-contract meeting to introduce the contractor to the client and all the other members of the project team

Indicative pre-contract meeting agenda

Ed Associates Ltd.

141 New Business Park, Anytown, AT51 1ST

Job no: EA – 123

Job title: NEW MUSIC CENTRE, ANYTOWN

AGENDA FOR PRE-CONTRACT MEETING

1 Introductions
- Appointments, personnel
- Roles and responsibilities

2 Contract
- Contract Sum
- Contract Documents
- Handover of production information
- Commencement and completion dates
- Insurances
- Bonds (if applicable)
- Standards and quality

3 Contractor's matters
- Possession
- Programme
- Health and Safety file and plan
- Site organisations, facilities and planning
- Security and protection
- Site restrictions
- Contractor's quality control policy and procedures
- Subcontractors and suppliers
- Statutory undertakers
- Planning and Building Regulation Approvals
- Overhead and underground services
- Temporary services
- Signboards

4 Clerk of works' matters *(if appointed)*
- Roles and duties
- Facilities
- Liaison
- Dayworks

5 Consultants' matters
- Contract Administrator
- Architectural
- Structural
- CDM-C/Health and Safety Plan
- Mechanical
- Electrical
- Others

6 Financial/Quantity Surveyor's matters *(if appointed)*
- Adjustments to tender figures
- Valuation procedures
- Cash flow
- Variations/remeasurement
- VAT

7 Communications and procedures
- Information requirements
- Distribution of information
- Valid instructions
- Lines of communication
- Dealing with queries
- Building Control notices
- Notices to adjoining owners/occupiers
- Samples

8 Any other business

9 Future meetings
- Pattern and proceedings
- Status of minutes
- Distribution of minutes

EA 123/ID May 2013
File refs: etc.

Figure 5.2 Indicative pre-contract meeting agenda

and endeavour to establish a good working relationship between them. In many instances they may not have worked together before and this may be the client's first project. If the meeting occurs too close to the date for commencement there may be insufficient time for the contractor to fully mobilise.

The contract administrator should issue an agenda, chair the meeting and issue concise and accurate notes as soon as possible after the meeting (see figure 5.2).

After the introductions, the first item on the agenda is to confirm the contract sum and the timescale. In many cases the contract sum will be the tender figure, but in other instances the tender sum may have been adjusted. This may be because of errors, additions or agreed post-tender cost reductions and should be set out as in the contract documents.

At this point the contract administrator should issue the documentation for construction in a pre-agreed format, such as architect's instructions forms, and agree the procedures for the issue of subsequent instructions. The issue of further information required by the contractor or, in the case of management contracts, any modifications to the information release schedule prepared in Stage 4 should also be agreed.

It is important that communications routes are agreed. The contractor should be advised that only formal instructions issued or countersigned by the contract administrator are valid and that all verbal instructions will be confirmed by a written instruction.

A difficulty often encountered is that the contractor will ask the client directly for instructions, or the client will issue instructions directly, particularly when the client is on site or lives there. Action should be taken to prevent this. Although the client may agree to something, they may not fully understand the time or cost implications. A simple example is if the client wants an electrical socket moving there is unlikely to be any time or cost implication if it hadn't been fixed but, if it had, the client will be liable for the time and cost in moving it. It may not always be possible to avoid this situation, but it should be reflected in the minutes to the pre-contract meeting by saying:

> *As far as possible it was agreed that any communication with the Contractor by the Employer should be via the*

> *architect/contract administrator. However, if this cannot be avoided then the architect/contract administrator should be made aware of any decisions immediately.*

Similarly the contract administrator should advise that:

> *Any work requiring subsequent covering up must not be done until the work has been inspected by the architect/contract administrator.*

Small project pre-contract meetings

When preparing for the pre-contract meeting a handy tip is that you can pre-draft most of the minutes as many of the items to be discussed are common to all projects and so can be adapted to suit the specific job. On small projects, the contractor is likely to have been through the process before, but that should not prevent the contract administrator from going through the full agenda as a reminder. In any event, particularly with small domestic projects, it should be remembered that the client may not have commissioned a building project before.

The minutes of the meeting should be distributed to all present and any other agreed parties as soon as possible after the meeting, with a note to say that if anyone has any comments or amendments then they should respond within seven days. If there are any amendments then the minutes should be reissued immediately after the seven days have expired as a true record.

In most cases, particularly with smaller projects, the pre-contract meeting will be held on the site rather than in the client's or consultant's offices. When the meeting is held on site, a site walk around should be conducted immediately after the formal meeting to resolve any queries arising in the meeting relating to access, site set-up and so on, so that any subsequent decisions can be recorded in the meeting minutes.

On larger projects the contract administrator should consider subsequently meeting with the contractor before the work starts to address any outstanding site matters from the pre-contract meeting.

What is the Construction Programme?

Construction Programmes describe the sequence in which tasks must be carried out so that a project (or part of a project) can be completed on time. Construction Programmes will identify dates and durations allocated to tasks. They can range from a simple bar (or Gantt) chart (see figure 5.3), through to a critical path analysis, where a sequence of critical tasks – upon which the overall duration of the programme is dependent – is highlighted. The critical path shows those activities that must be completed in a specific order so that the project can be completed on time. The critical path is usually shown as a series of lines and circles, with each circle representing an activity that needs to be completed and each line showing the relationship between two activities. The critical path will be the longest sequence of activities through the diagram, and will show how long a project is expected to take – assuming the scope does not change and everything goes according to plan.

Indicative bar/Gantt chart

Ed Associates Ltd.
141 New Business Park, Anytown, AT51 1ST

JOB No.	JOB TITLE									
123	**New Music Centre**									
	GANTT CHART/ CONSTRUCTION PROGRAMME									
	APRIL				JUNE				JULY	
Week no.	*1*	*2*	*3*	*4*	*5*	*6*	*7*	*8*	*9*	*10*
SET UP SITE										
EXCAVATION										
CONCRETE FOOTINGS										
BRICKWORK TO DAMP										
BEAM AND BLOCK FLOOR										
BRICKWORK TO PLATE										
ROOF STRUCTURE										
ROOF FINISHES										

Figure 5.3 Indicative bar/Gantt chart programme

For a Construction Programme to be effective, it must be used by the parties as a tool to help plan activities, monitor progress and identify where additional resources may be required.

The contractor's Construction Programme schedules all the construction activities based on their timescales, paying particular attention to: elements with long lead-in times; pre-contract works, such as demolition and site clearance; prefabricated elements; works outside the main contract, phasing and sectional completion.

Programmes can also incorporate the information release schedules to set out when the contract administrator and design team need to issue Technical Design information to the contractor in order for the works to progress. On design and build or management contracts, the programme should also indicate when information produced by the contractor's design team or specialist subcontractors should be issued to the contract administrator for comment and integration into the overall design.

The contract administrator should keep the Construction Programme under review, and report progress to the client on a regular basis, but not less frequently than at each site meeting.

The contractor's Construction Programme is not typically part of the contract documents and so is not enforceable under the Building Contract. The contract completion date, however, is enforceable and failure by the contractor to meet the completion date may lead to a claim by the client for liquidated damages. It should be noted that the completion date indicated on the contractor's Construction Programme may be earlier than the completion date entered in the Building Contract, to allow the contractor some float during the contract period. Indeed, it is not unusual for the contractor to produce two programmes: the formal one, which is issued to the client and the contract administrator, and the one to which they are working on site, which shows an earlier completion date.

How is the contractor instructed?

Formal instructions, historically known as architect's instructions, are issued by the contract administrator to authorise the contractor to carry out the duties associated with the contract. On design and build projects, only the client (the employer) or the employer's agent has the power to issue instructions. All instructions must be in writing, therefore a standard form is preferable to a letter (see figure 5.4). When issuing instructions the contract administrator needs to be aware that they can have a number of consequences, such as varying the design, increasing the amount of work and adding to the overall costs. Under the terms of the Building Contract, the contractor must comply with these instructions.

All instructions must be in writing and all verbal instructions should be confirmed by the contract administrator within seven days. If a verbal instruction is not confirmed by the contract administrator then the contractor should confirm it within the seven-day period, following which the contract administrator has a further seven days to dissent. If the contract administrator does not dissent then the instruction takes effect from the date of the contract administrator's notification. That said, if no one confirms and the contractor does the work, the instruction can be issued retrospectively, before the issue of the final certificate, *but this should be avoided.*

Instructions can be issued for a number of reasons, including:

- discrepancies
- issuing of further drawings and revisions
- expenditure of provisional sums
- variations in the works
- notification of defects arising either before or after Practical Completion
- confirmation of verbal instructions
- to open up work for inspection.

If the instruction varies the work the contract administrator should ideally have the work priced before issuing the instruction or the works described in the instruction should not commence until costs have been agreed.

All instructions should be numbered consecutively and should be worded to avoid the possibility of ambiguity in their interpretation.

The original instruction should be sent to the contractor, with duplicate copies sent to the client and the other design team members, including the cost consultant where employed, as agreed at the pre-contract meeting. They should be signed by the contract administrator before issue and can be issued in hard copy or, more commonly, by email as a PDF.

Contract administrator's instruction no. 1, which should preferably be handed over to the contractor at the pre-contract meeting, should confirm

Indicative contract administrator's instruction no. 1

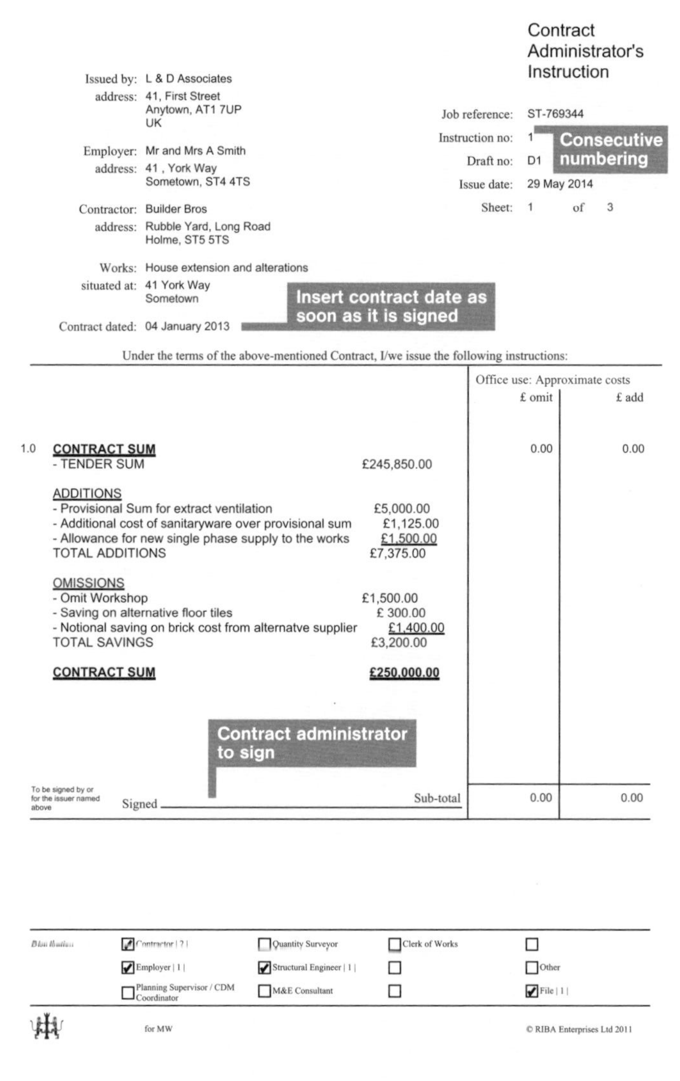

Contract Administrator's Instruction

Issued by: L & D Associates
address: 41, First Street
Anytown, AT1 7UP
UK

Employer: Mr and Mrs A Smith
address: 41 , York Way
Sometown, ST4 4TS

Contractor: Builder Bros
address: Rubble Yard, Long Road
Holme, ST5 5TS

Works: House extension and alterations
situated at: 41 York Way
Sometown

Contract dated: 04 January 2013

Job reference: ST-769344
Instruction no: 1
Draft no: D1
Issue date: 29 May 2014
Sheet: 1 of 3

Under the terms of the above-mentioned Contract, I/we issue the following instructions:

			Office use: Approximate costs £ omit	£ add
1.0	**CONTRACT SUM**		0.00	0.00
	- TENDER SUM	£245,850.00		
	ADDITIONS			
	- Provisional Sum for extract ventilation	£5,000.00		
	- Additional cost of sanitaryware over provisional sum	£1,125.00		
	- Allowance for new single phase supply to the works	£1,500.00		
	TOTAL ADDITIONS	£7,375.00		
	OMISSIONS			
	- Omit Workshop	£1,500.00		
	- Saving on alternative floor tiles	£ 300.00		
	- Notional saving on brick cost from alternatve supplier	£1,400.00		
	TOTAL SAVINGS	£3,200.00		
	CONTRACT SUM	**£250,000.00**		
	Sub-total		0.00	0.00

To be signed by or for the issuer named above — Signed ____________

- [x] Contractor | 2 |
- [ ] Quantity Surveyor
- [ ] Clerk of Works
- [ ]
- [x] Employer | 1 |
- [x] Structural Engineer | 1 |
- [ ]
- [ ] Other
- [ ] Planning Supervisor / CDM Coordinator
- [ ] M&E Consultant
- [ ]
- [x] File | 1 |

for MW

© RIBA Enterprises Ltd 2011

Figure 5.4 Indicative contract administrator's instruction no. 1

Indicative contract administrator's instruction no. 1 (*continued*)

Instruction continuation

Issued by: L & D Associates
address: 41, First Street
Anytown, AT1 7UP
UK

Job reference: ST-769344
Instruction no: 1
Draft no: D1
Issue date: 29 May 2014
Sheet: 2 of 3

Under the terms of the above-mentioned Contract, I/we issue the following instructions:

		Office use: Approximate costs £ omit	£ add
	Brought forward	0.00	0.00
2.0	**DRAWING ISSUE** Please carry out work in accordance with the drawings as listed on the attached Issue Sheet, two copies of which are enclosed. *Please cancel any previous copies of these drawings.*	0.00	0.00
3.0	**SPECIFICATION** Please carry out work in accordance with the specification (Revision A), two copies of which are enclosed. *Please destroy any previous copies.*	0.00	0.00
4.0	**STRUCTURAL ENGINEERING ISSUE** Please carry out work in accordance with the Structural Engineer's drawings and calculation sheets ref: 3040 as listed on the attached Issue Sheet, two copies of which are enclosed.	0.00	0.00
	Sub-total	0.00	0.00

Issue of information

Issue of information

To be signed by or for the issuer named above Signed ____________

Distribution

- [x] Contractor [2]
- [ ] Quantity Surveyor
- [ ] Clerk of Works
- [ ]
- [x] Employer [1]
- [x] Structural Engineer [1]
- [ ]
- [ ] Other
- [ ] Planning Supervisor / CDM Coordinator
- [ ] M&E Consultant
- [ ]
- [x] File [1]

for MW

(continued)

Indicative contract administrator's instruction no. 1 (*continued*)

Instruction continuation

Issued by: L & D Associates
address: 41, First Street
Anytown, AT1 7UP
UK

Job reference: ST-769344
Instruction no: 1
Draft no: D1
Issue date: 29 May 2014
Sheet: 3 of 3

Under the terms of the above-mentioned Contract, I/we issue the following instructions:

Omit all provisional sums until ready to place orders or expend monies

		Office use: Approximate costs £ omit	£ add
	Brought forward:	0.00	0.00
5.0	**PROVISIONAL SUMS (A54)** - Please omit from the contract the provisional sum for defined work: Additional work to foundations in the sum of £500.00. - Please omit from the contract the provisional sum for work by specialist subcontractor: security alarm in the sum of £1,400.00. - Please omit from the contract the provisional sum for contingencies in the sum of £5000.00. *- repeat for other provisional sums, dayworks, etc.*	6,900.00	0.00
6.0	*ADD ANY OTHER ITEMS/VARIATIONS AS REQUIRED*	0.00	0.00
6.1	**FLOOR TILES** Please omit specified floor tile in spec clause 32.006 and replace with Coral Tegula as agreed, saving accepted as item 1.0 above.	0.00	0.00
6.2	**BRICKS** Please omit bricks as specified in clause 25.009 at £800.00/1000 and substitute modular Hanson Butterley Wilnecote Red Rustic at a cost of £590/1000+OHP, saving in item 1.0 to be adjusted accordingly. Actual costs to be agreed.	0.00	0.00
To be signed by or for the issuer named above	Signed ________ Sub-total	6,900.00	0.00

Amount of Contract Sum/Tender Price	£	250,000.00
±Approximate value of previous issued Instructions	£	0.00
Sub-total	£	250,000.00
±Approximate value of this Instruction	£	-6,900.00
Approximate adjusted total	£	243,100.00

Distribution

☑ Contractor | 2 |	☐ Quantity Surveyor	☐ Clerk of Works	☐
☑ Employer | 1 |	☑ Structural Engineer | 1 |	☐	☐ Other
☐ Planning Supervisor / CDM Coordinator	☐ M&E Consultant	☐	☑ File | 1 |

for MW

the contract sum and issue further copies of the construction drawings and specification. If the drawings have been updated since the tender, change control would be required if there has been a variation to ensure that the cost and time implications of the change have been fully assessed. Within the contract sum there may have been a number of provisional sums and contingencies – these should be omitted from the Building Contract until such time as the relevant works are instructed.

What about the issue of instructions from other members of the design team?

During the course of the works on projects where other design team members are employed, there will inevitably be a requirement for them to issue instructions.

Design team instructions for small projects

On a small project the simplest method is for a design team member to draft their instruction and send it to the contract administrator to issue on one of their instructions. However, the contract administrator should note that the instruction comes from, say, the structural engineer, particularly if it is confirming a verbal instruction.

On larger projects it is often agreed that each consultant should issue their own instructions in a numbered sequence, but with a prefix as an identifier, eg SE/001 for the structural engineer. In such cases the contract administrator should either countersign the instruction or confirm the issue on their own instruction (as technically only the contract administrator is authorised to issue instructions and the contractor could dispute an instruction from another consultant). In this way a record of the instructions generated by each consultant can be kept together.

What are the tasks and procedures after work starts on site?

Once the contractor commences work on site it is the responsibility of the contract administrator to monitor the progress of the works and to authorise payments to the contractor.

Inspecting the works

The contract administrator's duty to inspect should have been defined for the project within the contract administrator's terms of appointment. On larger projects, where the contract administrator may not be responsible for the inspections, it may be other members of the project team who carry out the inspections, but the contract administrator should be informed of the details. Whoever carries out the inspections they should inspect the site and the progress of the works with reasonable skill and care. However, the term 'reasonable' does not mean that the inspection should go into every matter in detail nor does it guarantee that the inspection will reveal or prevent defective work. Indeed, it should be referred to as periodic inspection *not* supervision, as the latter has more onerous implications.

Periodic inspection and supervision

Periodic visits to site will cover visual monitoring of progress, but will not normally involve detailed checking of dimensions or testing of materials.

The contractor, on site, will *supervise* the work on a day-to-day basis and will be responsible for the proper carrying out and completion of construction works and for health and safety provisions on the site.

Periodic visits should be made as considered necessary and the timing and number should depend on the nature of the job – they are not exhaustive or continuous inspections. The contract administrator should make their own decisions on the most appropriate times to inspect and should not rely on the contractor for advice on when such inspections should be carried out. The contract administrator should allow sufficient time to carry out the checks properly and must not be waylaid or rushed by the contractor. However, experience shows that the degree of inspection will depend on the contract administrator's confidence in the contractor and that it is important to set up good lines of communication with the contractor.

Another factor which can have a bearing on the frequency of visits is how far the site is from the contract administrator's office, although the project size will also have an influence. Clearly, if it will take a few hours to get to site, it is not economic to carry out frequent inspections nor is it possible to visit the site at short notice should an issue arise. The contract administrator should take this into account when working out their fee for this stage. In any event, site visits should be made at least every two weeks for smaller projects, although monthly is more common on larger projects, particularly where there may be site-based design team members.

When the contract administrator is carrying out an inspection, the onus is on the contractor to provide safe access to the contract administrator. The contract administrator should always insist that safe access is provided, not only for their own safety, but also to check that the work is done properly. For example, the contractor should not be allowed to take down the scaffolding without the contract administrator having carried out a full inspection of the roof. However, the contract administrator should inspect the work in a timely manner to avoid delaying the dismantling of the scaffolding by the contractor.

The contract administrator should ensure that the contractor gives notice of when any work will be covered up, to enable the contract administrator to inspect the work before it is hidden.

Depending on what is stated in the Schedules of Services, inspections can comprise predictive visits, periodic inspections and spot checks. The contract administrator should prepare a list of all parts of the design that it is essential to check and, before each visit, review the drawings and check any previous notes.

The contract administrator should always make careful notes of inspections and write a short report or prepare a list. A checklist, adapted to the particular project, is a useful tool and acts a reminder when on site. It is good practice to have a standard form for recording visits (see figure 5.5), which can be used on site as an aide memoire. The following items should also be considered and noted:

- the quality of completed work
- the progress of the work in relation to the Construction Programme, including the start and completion dates of sections of the work
- queries from the contractor and answers given
- the weather conditions, where they may affect progress
- resolution of nonconformities, which are to be confirmed in writing
- any other factors which may affect the progress or quality of the work.

Indicative report form for predictive site visits

Ed Associates Ltd.
141 New Business Park, Anytown, AT51 1ST

Job no.: *EA123* Job title: *NEW MUSIC CENTRE*

SITE VISIT REPORT

Date:	*31nov12*	No. of visits scheduled	*24*
Visit by:	*ID*	Visit no.:	*10*

Purpose **Actions**

Inspect roof structure
Recheck brickwork and pad stones

Observed

Brickwork from last visit has now been rebuilt and is satisfactory
Pad stones now completed and installed
Steelwork 50% no comments at this stage.
Wrong insulation used in ceiling void – to be replaced – CHECK ON NEXT VISIT

Checked	Recorded
Samples	Photos
Verification of tests	
Vouchers	Video
Records	Other

Summary

Work properly executed ☐ Proceeding in workmanlike manner ☐

Materials properly stored and protected ☐ Progress to programme ☐

Figure 5.5 Indicative report form for predictive site visits

Indicative form for defective work

Ed Associates Ltd.
141 New Business Park, Anytown, AT51 1ST

Job no.: *EA123* Job title: *NEW MUSIC CENTRE*

DEFECTIVE WORK NOTED

Date	Item	Contractor notified	Value if deducted (£)	Date cleared
29jan12	*Existing manhole covered over under floor. Take up beam and block floor locally and raise to floor level as drawings. Make good floor and fit double-sealed cover as specification.*	*29jan12*	*0.00*	*16 feb12*

Figure 5.6 Indicative form for defective works

Report forms should be filled in during the site inspection and any defects noted (figure 5.6). At the end of the project these forms will build up a picture of any issues which have arisen.

Towards the end of the contract period the contractor will ask the contract administrator to prepare 'snagging lists' to assist the contractor to get the building ready for handover. This should be resisted as it is wholly the contractor's responsibility to have the building ready for the handover. Many contract administrators do still prepare snagging lists – this can be a time-consuming task, so if it becomes apparent that the building is not ready for handover then the contract administrator should terminate the inspection. However, it can still be useful for the contract administrator to go around with the contractor to point out specific items, but they should leave the obviously incomplete items to the contractor to complete.

What is snagging?

'Snagging' does not have an agreed meaning – it is a slang expression, not a contractual term. However, it is widely used to refer to the process of inspection necessary to compile a list of minor defects or omissions in building works for the contractor to complete, put right before Practical Completion or include with the Practical Completion certificate.

Depending on the nature of the project there may also be a requirement for inspections by other consultants, particularly where these are of a specialist nature and outside the detailed knowledge of the contract administrator. These should be reported in the same way as the contract administrator's inspections, with copies of the reports passed to the contract administrator.

In addition, on large projects a full-time resident consultant team or a clerk of works may be employed to monitor work on site. It is the duty of the clerk of works to assist the contract administrator by observing, inspecting, checking and reporting to the contract administrator on a regular basis and to keep a site diary. Their role might include:

- witnessing tests
- monitoring progress against the Construction Programme
- assessing whether the works comply with legal requirements, such as health and safety legislation
- assessing whether the works are being carried out in accordance with the contract documents, including taking measurements and samples if necessary
- monitoring site conditions and providing weekly reports to the contract administrator

Site inspections

Further guidance and detailed information on inspecting works can be found in the *RIBA Good Practice Guide: Inspecting Works by Nicholas Jamieson (November 2009).*

- attending construction progress meetings
- keeping records of progress, delays, weather conditions, details of information received, deliveries, health or safety issues or any other significant events.

What about Design Queries?

Design Queries can be raised by the contractor at any time, whether verbally during site inspections, by phone or email or via formal query sheets. Design Queries are typically either requests for information (RFI) or technical queries (TQ). Many contractors will produce formal query sheets to help them keep track of any responses. On receipt of a query sheet, the contract administrator should seek a response from the appropriate design team member as soon as possible, complete the query sheet, date it and return it to the contractor. Design Queries generated by the contractor can arise for a number of reasons, such as errors in or clarification on the information issued, matters arising from unexpected findings on site and requests for more information. Irrespective of the nature of the Design Query, the contract administrator should record and date their answer and issue an instruction or design change notice (see figure 4.15) as appropriate.

The contractor should also give notice of any extra costs to the contract administrator, who should sign and return them to the contractor if in agreement before formally confirming them on a contract administrator's instruction. However, if the contract administrator disagrees with notices these should be returned as 'not accepted'.

What about checking drawings?

In addition to responding to Design Queries, the contract administrator is required to check any specialist drawings prepared by subcontractors for the contractor's designed portion where applicable.

What are variations?

It is fully accepted that any contract, irrespective of size, is never carried out without any variations, whether additions or omissions. Variations may include alterations to the design, quantities, quality, working conditions and sequence of work. Variations are less likely to occur on design and build projects, assuming that the design and brief remain unchanged, but the client should be made aware that any changes they make could have an impact on the overall cost and timescale.

The contract administrator should keep the client fully aware of all variations which have a financial effect so there can be no grounds for complaint when the final account is being prepared. It is usually frustrating for the client to be faced with a bill for extras, whether generated by the client or the contract administrator.

A traditional contract should not only allow for expenditure on provisional sums, but also have a contingency sum to cover any unforeseen matters. The contingency sum should not be seen as a pot of money for paying for work not originally envisaged, it should be used only for genuinely unforeseen items. However, if there is contingency money remaining at a late stage in the project, when the risk of unforeseen matters arising is minimised, then it may be possible to use some of the money, but the contact administrator should still be cautious.

When the works give rise to a variation or additional work, it should ideally be priced before it is instructed, unless the delay could interfere with the regular progress of the works. In such cases the costs should be agreed as soon after the event as possible and the client notified accordingly.

Variations are often sources of dispute (either in valuing the variation, or in agreeing whether the work constitutes a variation at all) and can cost a lot of time and money during the course of a contract. While some variations are unavoidable, it is wise to minimise potential variations and subsequent claims by ensuring that uncertainties are eliminated before awarding the Building Contract.

Variations should always be issued in sufficient time for the work to be included within the building operation it affects to avoid the potential for delays. Whatever form the variation takes, the contract administrator should ensure that the agreed Change Control Procedures are complied with (see Stage 4, page 127).

Are there any limits on variations?

There are technically no limits on the number of variations with JCT contracts, but with the NEC3 and FIDIC forms of contract limits are placed on the extent of variations.

NEC3 (Option B) is a remeasurement and compensation form of contract in which the prices are based on a priced bill of quantities.

When an item is remeasured, if its quantity is different to that in the original bill and as a result the cost per unit changes, and the total remeasured quantity of that item multiplied by the original bill rate is greater than 0.5% of the total of the original bill, then a compensation event occurs, which then allows the unit price to be amended.

FIDIC forms of contract also put limitations on variations that can be instructed. If the value of the contract increases or decreases by more than 15% of the net contract sum (excluding provisional sums and dayworks) then the contract administrator can add or deduct from the contract sum a determined value upon consultation with the contractor, having due regard to their site expenses and other general overheads. Note that this 15% increase or decrease is not for a single item of work, but the total contract sum at completion.

What is the purpose of a site meeting?

Site meetings should be convened on a regular basis (usually at the same frequency as the valuations), such as on the second Thursday of each month or every four weeks. On traditional forms of contract it is usually the contract administrator who chairs the meeting, whereas on design and build and management contracts it is the contractor.

On larger projects the main contractor should chair their own technical meeting with the design team and the specialist subcontractors, often immediately prior to the main progress meeting, so that they can report the outcomes.

The meeting should only be attended by those persons involved in the project at the time. For example, once the ground works and structure are complete it is unnecessary for the structural engineer to attend.

The client or their representative is always welcome to attend – this is more common on larger projects. On smaller projects, particularly where there is not a professional client, they may not attend. In such instances the contract administrator should ensure that they speak to the client before the meeting to check on any actions from the previous meeting and to see if they have any problems or concerns. These can then be reported at the meeting. At the same time, or immediately after the meeting, they

Indicative site progress meeting agenda

Ed Associates Ltd.

141 New Business Park, Anytown, AT51 1ST

Job no.: EA123 *Job title: NEW MUSIC CENTRE*

AGENDA FOR SITE PROGRESS MEETING

ACTION

1 Minutes of last meeting not on the agenda

2 Statutory authorities

3 Contractor's report
- General report
- Subcontractors' meeting report
- Progress and programme
- Causes of delay
- Health and safety matters
- Information received since last meeting
- Information and drawings required
- Contract administrator's instructions required

4 Consultants' report
- Architect's report
- Structural works
- Mechanical works
- Electrical works
- CDM Coordinator
- Others

5 Financial/quantity surveyor's report

6 Design Queries

7 Contract completion date
- Likely delays and their effect
- Review of factors from previous meeting
- Factors for review at next meeting
- Revision to completion date
- Revisions required to programme

8 Any other business

9 Date of next meeting

File refs, etc.

Figure 5.7 Indicative site progress meeting agenda

can give them an update on progress in terms of:

- *time*: are there any changes to the completion date?
- *quality*: are there any issues with the materials and workmanship?
- *cost*: have there been any cost adjustments?

An agenda should be issued prior to the meeting and, to save time especially on larger projects, each party attending should be asked to submit written reports before the meeting (see figure 5.7).

The contract administrator may decide to precede the meeting with a site visit to get a prior indication of progress since the previous visit. In general terms the meeting should primarily be to receive reports and to agree actions to be taken, not to receive and answer queries. The latter may be dealt with in ad hoc meetings, but the contract administrator should still ensure that the details are recorded.

Site meetings for small projects

In the case of small projects it is often useful to include an agenda item for Design Queries, to make the meeting more effective.

Detailed and accurate notes of the meeting should be taken by the chairperson or their representative with a separate column to indicate who is to take any action. The notes of the meeting should be distributed to all present and any other parties as agreed as soon as possible after the meeting, with a note to say that if anyone has any comments or amendments then they should respond within seven days. If there are any amendments then the notes should be reissued immediately after the seven days has expired as a true record.

What are the tasks and procedures for payments?

One of the responsibilities of the contract administrator is to issue certificates for payments due to the contractor. Except on short-term projects, the contractor should be provided with regular instalments of money to enable them to finance the work. Payment of the sums is regulated by the contract administrator, whose duty is to certify what amount is due for payment at each instalment date.

What is an interim certificate?

Interim certificates provide a mechanism for the client to make payments to the contractor before the works are complete. The Housing Grants, Construction and Regeneration Act 1996 states that a party to a construction contract in excess of 45 days' duration is entitled to interim or stage payments.

The contract administrator must exercise reasonable skill and care when issuing certificates to ensure that the values reflect the value of the work that has been carried out. The contract administrator must be impartial and unbiased when certifying payments – they must be satisfied that the payments are for the correct amount as it is unfair to pay less, but it is also unwise to pay more. Should the contract administrator certify too much and the contractor defaults, the client will incur additional costs to complete the works – if the contract administrator has over certified, the client could hold them liable for that loss.

Certificates for payment are known as *interim certificates* or *certificates of progress payments*, depending on which form of contract is being used, but essentially they all do the same thing, which is to instruct the client to pay the contractor an agreed sum of money at a particular time.

The contractor should submit their valuation, usually immediately prior to the progress meeting, based on the value of all the work that has been completed to that date, including any variations. Once the total is agreed by the contract administrator the amount is entered on the interim certificate less any amounts already paid with 5% retention deducted during the contract period. Once issued, the client must honour the certificate within the period stipulated by the contract, usually 14 days.

Interim certificates should state the amount of retention. A statement should also be prepared showing the retention for nominated subcontractors, if there are any. On a staged contract the statement of retention should include relevant retentions for the different stages, eg 5%, 2.5% or 0%. Some contracts may require that the retention is kept in a separate bank account and that this is certified. In this case, the client will generally keep any interest paid on the account.

There may be provision to include the value of items that the contractor has not yet delivered to site or items under manufacture or stored at works awaiting delivery. This allows the contractor to order items in good time, without incurring unnecessary long-term expense, but it does put the client at some risk if the contractor or the supplier becomes insolvent.

On design and build projects, the amounts certified as payable may be based on the contract sum analysis prepared by the contractor as part of their tender. On small projects the valuation is usually prepared by the contractor or their estimator and submitted to the contact administrator for evaluation and certification. The contract administrator should ensure that it is in a form which can be evaluated to arrive at a fair and reasonable price. Often contractors will want to be paid at the end of a construction stage, such as when the foundations are in, but these will not always fall at regular intervals. A more efficient way of being able to evaluate the works is to have the contractor prepare a spreadsheet based on the priced schedule of work and for each valuation to proportion the amount of work done in each section and to list any variations. This provides a simple and efficient way for both parties to arrive at a fair valuation.

Once the valuation is agreed, the interim certificate can be completed, as shown in figure 5.8, with the retention and previous payments deducted to arrive at the amount due, exclusive of VAT. VAT is no longer shown on most certificates as VAT may not be applicable on some projects, such as new-build housing and listed buildings, where the onus for charging VAT rests with the contractor.

The completed certificate together with a copy of the valuation should be sent to the client with a covering letter instructing them to pay the contractor. However, for clients who are unable to claim back VAT, it is helpful to them if the VAT inclusive figure is stated on the covering letter. A suitable wording for the letter is shown in figure 5.9.

At the same time, the contract administrator should send a similar notification to the contractor with a copy of the certificate, and a copy of the valuation if it has been amended during evaluation, with a request to send a VAT invoice to the client by return for the amount due in the certificate. Under the terms of the Building Contract the client should pay on receipt of the certificate, but many clients need a VAT invoice in order to process the payment through their system.

On larger projects, where a cost consultant has been employed, the valuations are normally prepared by the cost consultant and the contractor

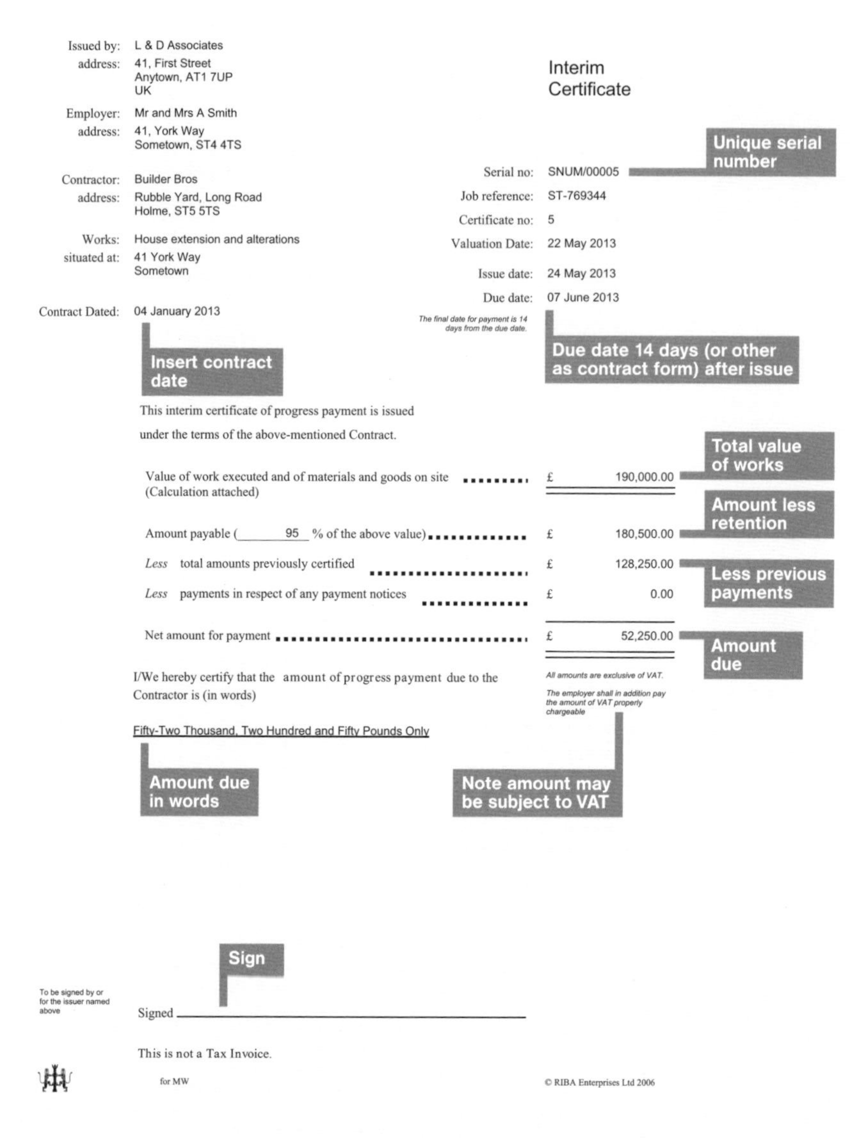
Indicative interim certificate

Issued by: L & D Associates
address: 41, First Street
Anytown, AT1 7UP
UK

Employer: Mr and Mrs A Smith
address: 41, York Way
Sometown, ST4 4TS

Contractor: Builder Bros
address: Rubble Yard, Long Road
Holme, ST5 5TS

Works: House extension and alterations
situated at: 41 York Way
Sometown

Contract Dated: 04 January 2013

Interim Certificate

Serial no: SNUM/00005
Job reference: ST-769344
Certificate no: 5
Valuation Date: 22 May 2013
Issue date: 24 May 2013
Due date: 07 June 2013

The final date for payment is 14 days from the due date.

This interim certificate of progress payment is issued under the terms of the above-mentioned Contract.

Value of work executed and of materials and goods on site (Calculation attached)	£	190,000.00
Amount payable (95 % of the above value)	£	180,500.00
Less total amounts previously certified	£	128,250.00
Less payments in respect of any payment notices	£	0.00
Net amount for payment	£	52,250.00

All amounts are exclusive of VAT.

The employer shall in addition pay the amount of VAT properly chargeable

I/We hereby certify that the amount of progress payment due to the Contractor is (in words)

Fifty-Two Thousand, Two Hundred and Fifty Pounds Only

To be signed by or for the issuer named above

Signed ______________________

This is not a Tax Invoice.

for MW

© RIBA Enterprises Ltd 2006

Figure 5.8 Indicative interim certificate

Indicative letter to client issuing certificates

Dear XXXX,

Re: PROJECT TITLE

Please find enclosed our Interim Certificate No. 5 in the sum of £52,250.00 plus VAT together with a copy of the valuation.

We have checked the valuation and we have no comments/the following comments:

- [Add comments]
-
-

On this basis the VAT inclusive sum of £62,700.00 is due to the Contractor within 14 days.

Please do not hesitate to contact me if you have any queries or need any further information.

Yours sincerely

Figure 5.9 Indicative letter to client issuing certificates

together and a figure agreed. The cost consultant will then submit the agreed valuation to the contract administrator, who will process the certificate in the usual way.

The contract administrator (or cost consultant where appointed) should monitor the costs on a regular basis and give the client regular cost updates to keep them informed of any variations in the overall cost. A professional client may only be interested in the contract sum, but a smaller client may be looking at the overall cost so the reports may need to include fees and VAT.

What are dayworks?

Daywork is a means by which a contractor is paid for specifically instructed work on the basis of the cost of labour, materials and plant plus a mark-up for overheads and profit when work cannot be priced in the normal way. Daywork may be applied, for example, when unforeseen circumstances are encountered during the works or when work is instructed for which there are no comparative rates in a bill of quantities or schedule of rates.

Most contracts contain clauses that provide a method of evaluating variations, additional work and instructions by using existing contract rates and prices, but NEC contracts are based on the contractor quoting for the work (see page 168).

Daywork rates can be priced in one of two ways: by either a percentage addition, where a percentage is added for overheads, profit and incidental costs; or by the use of all-inclusive rates. All-inclusive dayworks rates are quoted at tender stage as part of the preliminaries and incorporated into the contract documents. These include an allowance for overheads and profit, either fixed for the period of the contract or, in the case of contract conditions that are index linked, subject to an inflation allowance.

What if the works are delayed?

Where works are delayed or will be delayed beyond completion and the cause of the delay is beyond the control of the contractor, including the impact of variations, the contractor may be entitled to an extension of time and, in certain circumstances, additional costs for loss and expense. Equally, the contractor may only be entitled to an extension of time with no costs attached such as in the case of delays caused by the contractor's own domestic subcontractors.

If the contractor fails to complete the works by the date set in the Building Contract the client has the right, under the terms of the contract, to deduction of liquidated and ascertained damages, which is a pre-agreed measure of compensation. However, a number of factors may have had an impact on the delay – the contract administrator will need to assess these factors and, if appropriate, issue an extension of time certificate.

An extension of time allows the contract administrator to change the date for completion of the works or a part of the works if the contractor is delayed by certain specified events, which releases the contractor from the need to pay liquidated and ascertained damages. Conversely it also retains the right of the client to deduct or withhold liquidated damages.

The contract administrator must act lawfully, rationally and methodically to decide whether the cause of delay is a 'relevant event' qualifying for an extension to the contract period. The contract administrator has a duty to use a calculated approach, not an impressionistic one.

To assess whether a cause of the delay constitutes a relevant event will depend on it falling into one of the following categories:

- variations to the design and detail
- instructions issued by the contract administrator
- the implications of any discrepancies between the drawings, specifications, etc.
- suspension of the works by the contractor
- loss or damage by the perils noted in the Building Contract
- delay in possession of the site or part of the site
- delays as a result of opening up the works as directed by the contract administrator *unless the works or the materials are found to be faulty*
- any actions by the client, consultant team or others employed by the client which hinder the regular progress of the works
- *exceptionally* inclement weather
- civil commotion and terrorism threats
- force majeure/'Acts of God'.

The award of an extension of time is solely the responsibility of the contract administrator, who must assess the time lost by the contractor so that their liability for damages is the same as if the disruption had not occurred. The contract administrator must remain firmly independent of the client in this respect.

What should the contract administrator do about delays?

Under the terms of the Building Contract, when it becomes reasonably apparent that the works are being or are likely to become delayed, the contractor should give written notice to the contract administrator of the delay and provide supplementary information to enable the contract administrator to make an assessment (see figure 5.10). However, having made a claim for lost time, the contractor is still under an obligation to use their best endeavours to make up for this lost time, even if the fault is not of their making.

The supplementary information required will be stated in the contract form, but will include the material circumstances and the cause or causes of the delay and details of any 'relevant events'.

The contract administrator should assess and make the award as soon as is reasonably practical, but within 12 weeks of the receipt of the detailed information and certainly no later than the completion date.

On NEC3 contracts, delays are included under the term 'compensation events', which are any matters that could affect time, cost or quality. There is an obligation on both parties to give early warnings of compensation events. Indeed, where a compensation event arises from the contract administrator issuing an instruction, or changing a previous one, the onus is on the contract administrator to notify the contractor of the compensation event.

If events occur during the course of the works that cause the completion of the works to be delayed then these may be compensation events. Compensation events will normally result in additional payment being made to the contractor and may result in the adjustment of the completion date or key dates.

NEC3 contracts limit compensation events to those, and only those, identified in the contract. If an event is not identified in the contract as being a compensation event then no claim should be submitted, even if there has been a delay. The contract prevents the parties circumventing the contract by making a claim for damages at common law.

The NEC3 assessment system is more formalised than in JCT contracts.

- If a compensation event occurs, the contractor must notify the contract administrator within eight weeks of the event or it is time barred.
- The contract administrator must respond within one week – if the contract administrator does not respond within two weeks it is deemed that the delay is accepted.
- The contractor should then quote for the compensation event within three weeks.
- The contract administrator must respond to the quotation within two weeks.
- The contractor then has three weeks to revise the quotation if required.
- When agreement has been reached, any changes to the contract are implemented.

Unlike in JCT contracts, the assessment of the delay caused by a compensation event is final and cannot be revised later if proved to be incorrect.

NEC3 contracts also make provision for early warning procedures. Both parties must give early warning of anything that may delay the works, or increase costs.

NEC3 contracts

Further details for assessing compensation events in NEC contracts can be found in section 9 of the *RIBA Good Practice Guide: Extensions of Time* by Gillian Birkby, Albert Ponte and Frances Alderson (June 2008).

In the past it was commonplace for the contract administrator to delay awarding an extension of time and they would wait and see whether the contractor could make up the time or by how much they could reduce the delay. This is now unnecessary as most forms of contract allow the contract administrator a period of 12 weeks following Practical Completion in which to reassess any extensions granted and to make adjustments if necessary. This may mean extending, or even reducing, extensions already granted. Note, however, that it is best to leave the full reassessment of all extensions of time until after completion, when the full picture is known and the contract administrator can carry out a complete review, as some forms will only allow one reassessment.

How should the contract administrator assess extensions of time?

In response to the information supplied by the contractor, the contract administrator needs to build up a full picture of the situation as soon as possible. Pre-existing records should be consulted, such as:

- site progress meeting minutes, where actual or potential delays should have been recorded
- records of site delays noted
- contractor's reports
- the Construction Programme
- method statements
- daywork sheets
- clerk of works' reports, where one is appointed
- other team members' records
- the as-built programme, if assessing after Practical Completion
- the contract administrator's own observations and site notes.

Indicative form for recording site delays

Ed Associates Ltd.
141 New Business Park, Anytown, AT51 1ST

Job no.: *EA123* Job title: *NEW MUSIC CENTRE*

RECORD OF SITE DELAYS NOTED

Date	Delay	Item	Reason	Observed by
16jan12	*5 days*	*No bricklaying*	*Heavy rain*	*Contractor*
12feb12	*2 weeks*	*Windows delayed*	*Delay by S/C*	*CA*

Figure 5.10 Indicative form for recording site delays

Instructions for variations should be issued in sufficient time for the work to be included within the building operation it affects. Therefore, the fact that a variation has been instructed will not necessarily mean that it was the cause of an operation being delayed. The objective for the contract administrator is to identify when a variation was executed and what impact it had on the building operation, based on the physical and material content of the variation.

When dealing with extensions of time, the contract administrator should remember the following:

- Respond to each and every proper notice of delay from the contractor – at least this is evidence that the claim has been considered.
- When awarding extensions of time, do so only for the causes specified in the contract. State the causes, but do not apportion. Keep full records in case the award is contested.
- Comply strictly with the procedural rules.

- Observe the timescale if one is stated in the contract. If none is stated, act within a reasonable time.
- Form an opinion which is fair and reasonable in the light of the information available at the time.

Acknowledging requests from the contractor

The contract administrator should acknowledge requests for extensions of time with the following response:

> *Thank you for your notification that the regular progress of the work has been materially affected and that this may give rise to an application for reimbursement of direct loss and expense under Clause [xxx] of JCT [xxx] [or alternative as specific to the Building Contract].*
>
> *Would you please state which of the list of events you consider relevant in this instance, and forward such information that you consider will reasonably enable us to form an opinion, including your reasons as to why you cannot recover the cost under any other condition.*
>
> *Should we be of the opinion that the delay is covered by Clause [xxx] you will be required to provide me [or the quantity surveyor] with such details as are reasonably necessary to ascertain the direct loss and/or expense.*

Once the contract administrator has assessed the claim, both the contractor and the client should be notified of the decision by issuing a suitable certificate, such as the one shown in figure 5.11.

On design and build forms the term 'extension of time certificate' is replaced by 'notice of revision to completion date', but the meaning and processes are the same.

When assessing any claims for extensions of time the contract administrator should be aware that the contractor may make more than one claim at a time, thus the evaluation process is more complex as some claims may attract costs and others not. These are known as concurrent delays (see below).

Indicative extension of time certificate

Notification of an
Extension of Time

Issued by: L & D Associates
address: 41, First Street
Anytown, AT1 7UP

Employer: Mr and Mrs A Smith
address: 41, York Way
Sometown, ST4 4TS

Job reference: ST-769344

Contractor: Builder Bros
Address: Rubble Yard, Long Road
Holme, ST5 5TS

Notification no: 1

Issue date: 29 May 2014

Works: House extension and alterations
situated at: 41 York Way
Sometown

Contract dated: 04 January 2013

Under the terms of the above-mentioned Contract,

I/we hereby give notice that the time for completion of the Works

is extended beyond the Date for Completion stated in the Contract or any later date previously fixed so as to expire on

19 July 2013

Date to which the completion date is extended to after evaluation

for the following reasons:

- Delays in starting work on site by the employer's kitchen fitter.

Include reasons for delay

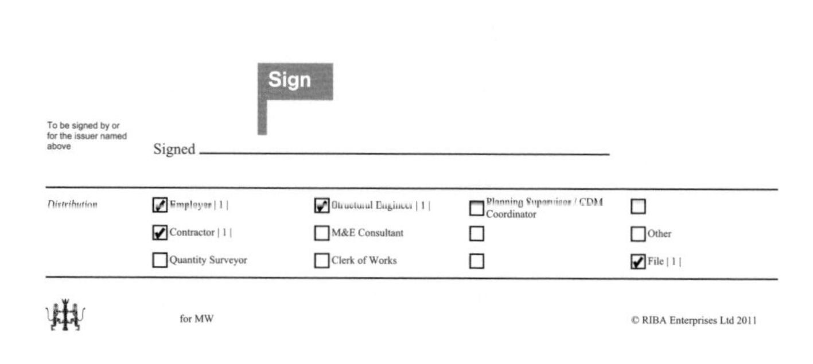
Sign

To be signed by or for the issuer named above

Signed ____________________

Distribution				
	☑ Employer | 1 |	☑ Structural Engineer | 1 |	☐ Planning Supervisor / CDM Coordinator	☐
	☑ Contractor | 1 |	☐ M&E Consultant	☐	☐ Other
	☐ Quantity Surveyor	☐ Clerk of Works	☐	☑ File | 1 |

for MW

© RIBA Enterprises Ltd 2011

Figure 5.11 Indicative extension of time certificate

Indicative record of contractor's claims

Ed Associates Ltd.
141 New Business Park, Anytown, AT51 1ST

Job no.: *EA123* Job title: *NEW MUSIC CENTRE*

RECORD OF CONTRACTOR'S CLAIMS

Date	Clause no.	Item	Time/amount Requested	Time/amount Allowed	Date for decision	Date awarded
19jan12		*Exceptionally inclement weather*	*5 days*	*3 days*	*22jan14*	*20jan14*
17feb12		*Delay in window installation*	*2 weeks*	*0*	*20feb14*	*N/A*

Figure 5.12 Indicative record of contractor's claims

To assist in such cases the contract administrator should keep a record of all claims arising and the decisions made in a suitable format, such as the one indicated in figure 5.12.

Further references to the assessment of extensions of time can be found in The *RIBA Good Practice Guide: Extensions of Time* by Gillian Birkby, Albert Ponte and Frances Alderson (June 2008).

What is loss and expense?

At the same time as an extension of time application the contractor can also make a claim for direct loss and expense for some of the contract provisions – the terms of the Building Contract should be consulted to ascertain the actual provisions. Once the contract administrator has confirmed that the application is valid and that sufficient information has been provided, their assessment of loss and expense must be based on the following:

- There must be an actual direct loss and the costs claimed must not allow the contractor to profit as a result of the disruption.
- The works must have been materially affected.
- Disturbance to the planned work must have already occurred.
- Payment for the work cannot be made under any other contractual provision.

The awarding of an extension of time and any associated loss and expense effectively puts the contractor back in the position they would have been in had the disruption not occurred.

What are concurrent delays?

Concurrent delay is a complex term, the legal status of which remains unclear. It refers to the situation where more than one event occurs at the same time, but where not all of those events enable the contractor to claim an extension of time or to claim loss and expense, therefore the contract administrator will have to assess the balance between the individual elements. Increasingly, such analyses are undertaken by experts who specialise in delay analysis, particularly on larger projects.

What can be the effect of late information?

The contract administrator must ensure that any information issued to the contractor during the course of Stage 5 should be timely and meet the contractor's information required schedule. Failure to do so may not only have an impact on the Construction Programme but may expose the client to a claim for an extension of time and additional costs from the contractor on the grounds of late information.

The contract administrator must avoid issuing information late as it will put them in a difficult situation. In such circumstances, the contract administrator must not be influenced improperly by any other party, and must continue to act professionally and fairly.

Conflicts of interest can arise where the contract administrator is also the designer. Part of the contract administration role is to manage errors and omissions within the design information. In such a situation it may be difficult for the contract administrator to act impartially, as decisions that they make as contract administrator may well have implications for their role as designer.

What if the contractor does not complete in time?

The date(s) for completion of construction works will have been set out in the contract particulars. However, if the date(s) for Practical Completion passes before the works are complete and the contract administrator has not granted any extensions of time, a Practical Completion certificate cannot be issued on the due date.

If the contractor is responsible for the delay and the contract administrator has not issued an extension of time, the client may be entitled to claim liquidated and ascertained damages at the rate set out in the contract particulars.

Some contracts, such as the JCT series, require that the contract administrator issues the contractor with a certificate of non-completion before the client can claim liquidated and ascertained damages. This gives formal written notice to the contractor that they have failed to complete the works described in the Building Contract by the completion date or other agreed date.

The contract administrator should have given due consideration to any applications for extension of time before issuing a certificate of non-completion. If there are subsequent extensions of time that result in the completion date being adjusted, and the contractor then fails to meet this adjusted date, a new certificate of non-completion must be issued.

On contracts with sectional completion, separate certificates of non-completion must be issued for each section that is not completed by the required date.

The client is entitled to deduct liquidated and ascertained damages from any subsequent payments due to the contractor, provided that a notice has been issued setting out the basis of the calculation. Note that this entitlement does not affect the contract administrator's duty to assess interim certificates which should be assessed as normal, with the certificate stating the full value of the work done. The onus is on the client to deduct the liquidated damages, not on the contract administrator

Indicative notice of non-completion

Notice of
Non-Completion

DB

Employer's Agent: ED Associates
address: 141, New Business Park
Anytown, AT51 1ST

From Employer: The Governors of AGood School
address: AGood School, London Road
Anytown, AT1 1TA

To Contractor: Building and Design Ltd
address: BD House, Main Road
Anytown, AT1 1AT

Works: New Music Centre
situated at: AGood School
London Road
Anytown

Contract dated: 28 June 2012

Job reference: EA - 5210

Notice number: 1

Date of issue: 29 May 2014

Under the terms of the above-mentioned Contract, notice is given that:

Complete if appropriate

the Works

~~Section no.~~ ____________________ ~~of the Works~~

Failed to be completed by the Completion Date of:

14 August 2013

Date set for Practical Completion or revised date through extension of time

To be signed by or for the Employer

Signed ____________________

for DB

Figure 5.13 Indicative notice of non-completion (design and build project option)

to reduce the certificate accordingly. The deduction can be challenged by the contractor if the procedures and the notice periods set out in the contract have not been followed.

It should be noted, however, that some forms of contract do not require a certificate of non-completion to be issued, but it is considered best practice to issue one on a standard form appropriate to the form of contract being used, such as the JCT design and build form (figure 5.13). (On design and build contracts the term 'certificate of non-completion' is replaced by 'notice of non-completion', but the meaning and processes are the same.)

What happens at Practical Completion?

When the project is nearing completion, a well-organised, efficient and effective transfer of information to the client is essential to ensure the successful handover and operation of the building, in accordance with the Handover Strategy. The period for the preparation for handover at Stage 6 precedes Practical Completion, which occurs at the end of Stage 5. The tasks and procedures relating to the preparation for handover will be dealt with in the next chapter.

What is Practical Completion?

There is no absolute legal definition of Practical Completion and case law is very complex. It has been defined as completion except for a few minor items when there are no apparent visible defects. From a practical point of view it may be acceptable to hand over the building if there are some cosmetic defects, but not if, say, the heating is not working.

Therefore, while the works may achieve Practical Completion despite the existence of latent defects (which are not known about), it should not be certified if there are patent defects. The contract administrator should be aware that patent and latent defects (see page 201 in Stage 6) are subject to different rules at law, which may be expressed in contracts, and this can cause confusion. However, there is some debate about whether Practical Completion can be certified where very minor items, which do not affect the client occupying the building, remain incomplete.

The contract administrator should certify Practical Completion when they consider that all the works described in the contract have been carried out (this is referred to as 'substantial completion' on some forms of contract). The issue of the certificate sets the date when the client can occupy the building and takes over responsibility for insurances. Certifying Practical Completion also has the effect of ending the contractor's liability for liquidated damages, releasing half of the retention monies and starting the rectification period.

Practical Completion in non-traditional contracts

On construction management contracts, separate certificates of Practical Completion should be issued for each trade contract. Once certificates for all the trade contracts have been issued, the contact administrator should issue a certificate for project or sectional completion. The same is true on management contracts, where each works contract must be certified individually.

Practical Completion is not a term recognised in PPC 2000 and other partnering contracts which simply refer to 'completion'. This is a far more open-ended term and it can put the contract administrator in a difficult position in deciding when the project becomes useable by the client. In such an instance the project may be deemed complete when the building is ready for the intended use by the client, whether it is for immediate use, such as fitting-out or actual occupation by the end users and this can be done in a safe manner without affecting warranties.

Often the contractor and the client are keen for the contract administrator to issue the certificate when more than minimal defects are outstanding. The contract administrator should be wary of bowing to either as the issue of the Practical Completion certificate could make them liable for subsequent problems, claims for which may arise from either party. This may affect a liquidated damages calculation or the premature release of retention monies, with no guarantees of when the works will be completed.

If the contract administrator has no option but to certify Practical Completion even though the works are not complete, they should inform the client in writing of the potential problems of doing so and obtain written consent from the client to do so. However, before issuing the certificate the contract administrator should also obtain a written agreement from the contractor that they will complete the works and rectify any defects.

What if the building is handed over in stages?

On large projects it is common for the Building Contract to provide for sectional completion, whereby different completion dates are set for different sections of the works. This allows the client to take possession of completed parts while construction continues on others.

Sectional completion is pre-planned and should be defined in the contract documents. The extent of each section must be clearly stated and the liquidated damages and the amount of retention that will be released must be specified for each section.

Sectional completion follows the usual handover procedures (see figure 5.14). However, it may exclude mechanical and electrical service systems, which may be reliant on total completion before they can be properly tested and commissioned.

Is partial possession different from sectional completion?

When the building is nearing completion the client or tenants may want to take possession of part of the building even if the works are ongoing or there are defects that have not been rectified. If this has not been pre-programmed as a sectional completion, the contract administrator may be able to arrange for the contractor to agree to partial possession.

The effect of partial possession is that any part for which partial possession is given is deemed to have achieved Practical Completion and half of the retention for that part must be released. The rectification period will begin for that part and liquidated damages will be reduced proportionally.

As with a single handover, the client will then be responsible for that part and should take out insurance cover for it.

The contractor is not obliged to allow partial possession (although permission cannot be unreasonably withheld) and may not wish to, particularly if it will hinder the completion of the remaining part and cost more as a result.

An alternative arrangement might be for the client to 'use or occupy' the works. This is different from possession in that the client does not have sole use of the works, and the contractor remains in possession, with

Indicative sectional completion certificate

Statement of
Section Completion

DB

Employer's Agent: ED Associates
address: 141, New Business Park
Anytown, AT51 1ST

From Employer: The Governors of AGood School
address: AGood School, London Road
Anytown, AT1 1TA

To Contractor: Building and Design Ltd
address: BD House, Main Road
Anytown, AT1 1AT

Works: New Music Centre
situated at: AGood School
London Road
Anytown

Contract dated: 28 June 2012

Job reference: EA - 5210

Statement number: 1

Date of issue: 29 May 2014

Under the terms of the above-mentioned Contract,

Insert section number

Section no. 1 of the Works

achieved Practical Completion

on 15 April 2013

Insert completion date of section

The Rectification Period in respect of the Section

is 12 Months

Insert period from contract

Sign

To be signed by or for the Employer

Signed ______________________

DB

Figure 5.14 Indicative sectional completion certificate

responsibility for insurance and so on. However, care should be taken with this option because of the possibility of the client occupying areas adjacent to, or only accessible through, the ongoing construction works.

Note that NEC3 has a different description of partial possession as 'taking over' the works.

What are the tasks and procedures at Practical Completion?

The tasks and procedures leading up to Practical Completion and the assembly of the necessary documentation ready for handover are dealt with at Stage 6.

The main contractor should give the contract administrator due notice that the building will, in their opinion, be ready for Practical Completion by a certain date. During this time the contract administrator should be carrying out their regular inspections to assess whether this is likely to be the case or not. The contract administrator should not, as explained earlier, provide snagging lists for the contractor as this is the responsibility of the contractor.

The pre-handover meeting (see page 197: Stage 6) should be arranged, at which the procedures and date for handover should be agreed.

By the date agreed for the Practical Completion meeting the contract administrator should have arranged for any other consultants to carry out their own inspections and to provide lists of any remaining defective and outstanding items to accompany the certificate.

The Practical Completion meeting should be attended by the client, and other consultants should attend in case they are required to answer queries. There should be a walk-round of every room in the building with the contract administrator making notes of defects as required. Once all parties are satisfied that the works are substantially complete and suitable for occupation, the client should accept the building and the keys should be handed over together with the maintenance and other documentation. Meter readings should be taken as necessary and the contractor should leave the site clear of all plant, etc.

Note that in the case of larger, more complex buildings, handover may occur over a period of a week, so that every room can be inspected before the final handover meeting.

Flow chart for tasks and procedures at Practical Completion

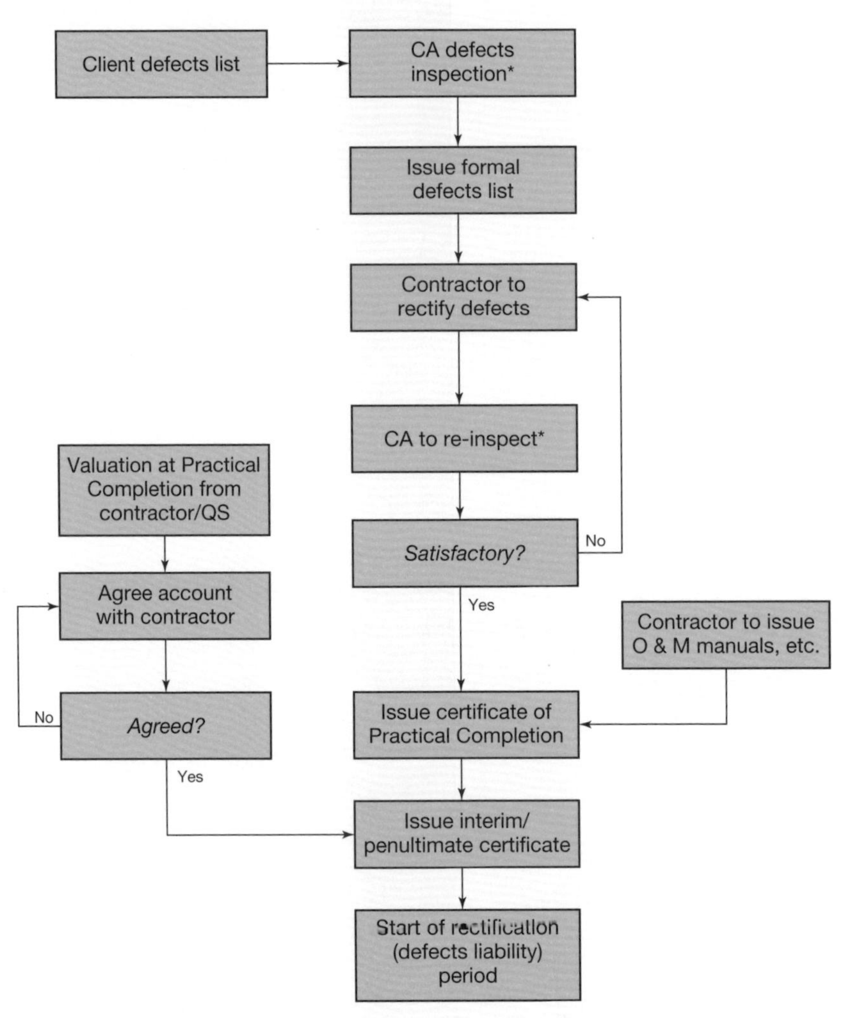

Note: In the case of sectional completion these actions should be undertaken for *each* section.
* With other consultants as appropriate
CA = contract administrator

Figure 5.15 Flow chart for tasks and procedures at Practical Completion

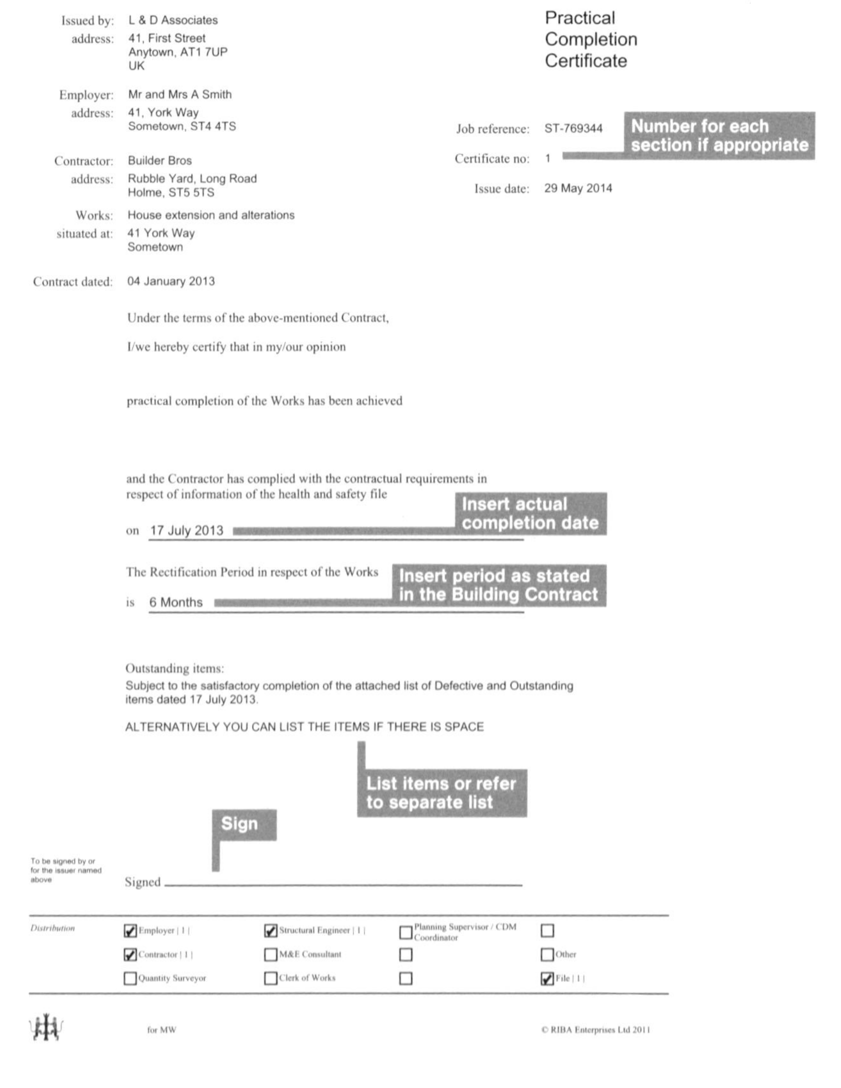

Indicative Practical Completion certificate

Practical Completion Certificate

Issued by: L & D Associates
address: 41, First Street
Anytown, AT1 7UP
UK

Employer: Mr and Mrs A Smith
address: 41, York Way
Sometown, ST4 4TS

Contractor: Builder Bros
address: Rubble Yard, Long Road
Holme, ST5 5TS

Works: House extension and alterations
situated at: 41 York Way
Sometown

Contract dated: 04 January 2013

Job reference: ST-769344
Certificate no: 1
Issue date: 29 May 2014

Under the terms of the above-mentioned Contract,

I/we hereby certify that in my/our opinion

practical completion of the Works has been achieved

and the Contractor has complied with the contractual requirements in respect of information of the health and safety file

on 17 July 2013

The Rectification Period in respect of the Works

is 6 Months

Outstanding items:
Subject to the satisfactory completion of the attached list of Defective and Outstanding items dated 17 July 2013.

ALTERNATIVELY YOU CAN LIST THE ITEMS IF THERE IS SPACE

To be signed by or for the issuer named above

Signed

Distribution

☑ Employer [1]	☑ Structural Engineer [1]	☐ Planning Supervisor / CDM Coordinator	☐
☑ Contractor [1]	☐ M&E Consultant	☐	☐ Other
☐ Quantity Surveyor	☐ Clerk of Works	☐	☑ File [1]

for MW

© RIBA Enterprises Ltd 2011

Figure 5.16 Indicative Practical Completion certificate

What are the tasks and procedures for issuing the Practical Completion certificate?

Once the building has been officially handed over, the contract administrator should assemble their list of defective and outstanding items at Practical Completion, including any items received from other consultants. The Practical Completion certificate appropriate to the form

Indicative letter to contractor following Practical Completion

Dear XXXX,

Re: PROJECT TITLE

Further to our meeting yesterday I would confirm that Practical Completion has been achieved and I enclose the Certificate of Practical Completion together with a list of minor defective and outstanding items which require your attention. Will you please confirm when the work is ready for re-inspection.

The Rectification Period will end on Friday 17 January 2014 when we will carry out an inspection but, in the meantime, following the receipt of your Valuation No. 8, this has now been evaluated and we have a number of comments in respect of the additional works. For the moment, we have omitted the extra cost of the wiring as we have deemed that this was part of the original works but should you wish to discuss this further and provide justification then we may be prepared to reconsider this.

I therefore enclose your copy of our Interim Certificate No. 8 in the sum of £14,576.25 plus VAT which is now being sent to the Employer for payment. This also releases half of the retention monies.

Please do not hesitate to contact me if you have any queries or need any further information but I must finally say that I was very pleased with the quality of the work. Well done.

Please do not hesitate to contact me if you have any queries or need any further information.

Yours sincerely

Figure 5.17 Indicative letter to contractor following Practical Completion

of contract used should be completed with the completion date and the date for the end of the rectification period (see figures 5.15 and 5.16).

There may be provision on the certificate to list the defects, but it is more practical to attach the list and add a statement to the form such as '*subject to the satisfactory completion of the attached list of defective and outstanding items dated [xxxx]*'.

Contractor notification and advice

Original certificates signed by the contract administrator should be issued to the contractor together with the list of minor items to be immediately rectified, with copies sent to the client and the other consultants.

Indicative letter to client following Practical Completion

Dear XXXX,

Re: PROJECT TITLE

Further to our meeting on 17 July 2013 I am pleased to advise that I can confirm that Practical Completion had been achieved and that you can move into the property. You will of course have to take over the insurance of the building formally so please advise me when this has been done. I enclose your copy of the Certificate of Practical Completion together with the list of minor defective and outstanding items which have to be completed by the contractor.

The Rectification Period will end on Friday 17 January 2014 following which I will carry out an inspection to identify any items which are attributable to the contractor. To assist in this respect I would be grateful if you could prepare a list of items which could be attended to at the end of the defects period but if anything is more urgent then please contact me directly and I will contact the contractor and get him to carry out the work as necessary.

I have asked the Contractor to complete the work as soon as possible but please do not hesitate to contact me if you have any queries or need any further information. In the meantime I hope that you will enjoy using the new spaces which have been created.

Yours sincerely

Figure 5.18 Indicative letter to client following Practical Completion

The certificate should be issued with a letter, such as that shown in figure 5.17, outlining the contractor's responsibilities during the rectification period.

Client notification and advice

The client should be advised of their responsibility to insure the building. It is prudent also to advise the client of what actions they should take if there are any matters relating to defects matters which arise before the end of the rectification period. A suitable form of words is noted in figure 5.18.

The tasks and procedures assume that there is a single handover, but in the case of sectional completion or partial possession, these tasks should be undertaken for *each* section.

Chapter summary 5

This chapter has demonstrated the tasks and procedures to be undertaken by the contract administrator during the construction process leading up to the Practical Completion of the works, when the client is able to take possession.

Other members of the project team will be undertaking other roles during this period, but the contract administrator is responsible for coordinating their work to ensure the satisfactory completion of the building.

Throughout this process the contract administrator must be methodical and thorough in the procedures to be undertaken, and must keep adequate records to provide an audit trail should any subsequent problems arise.

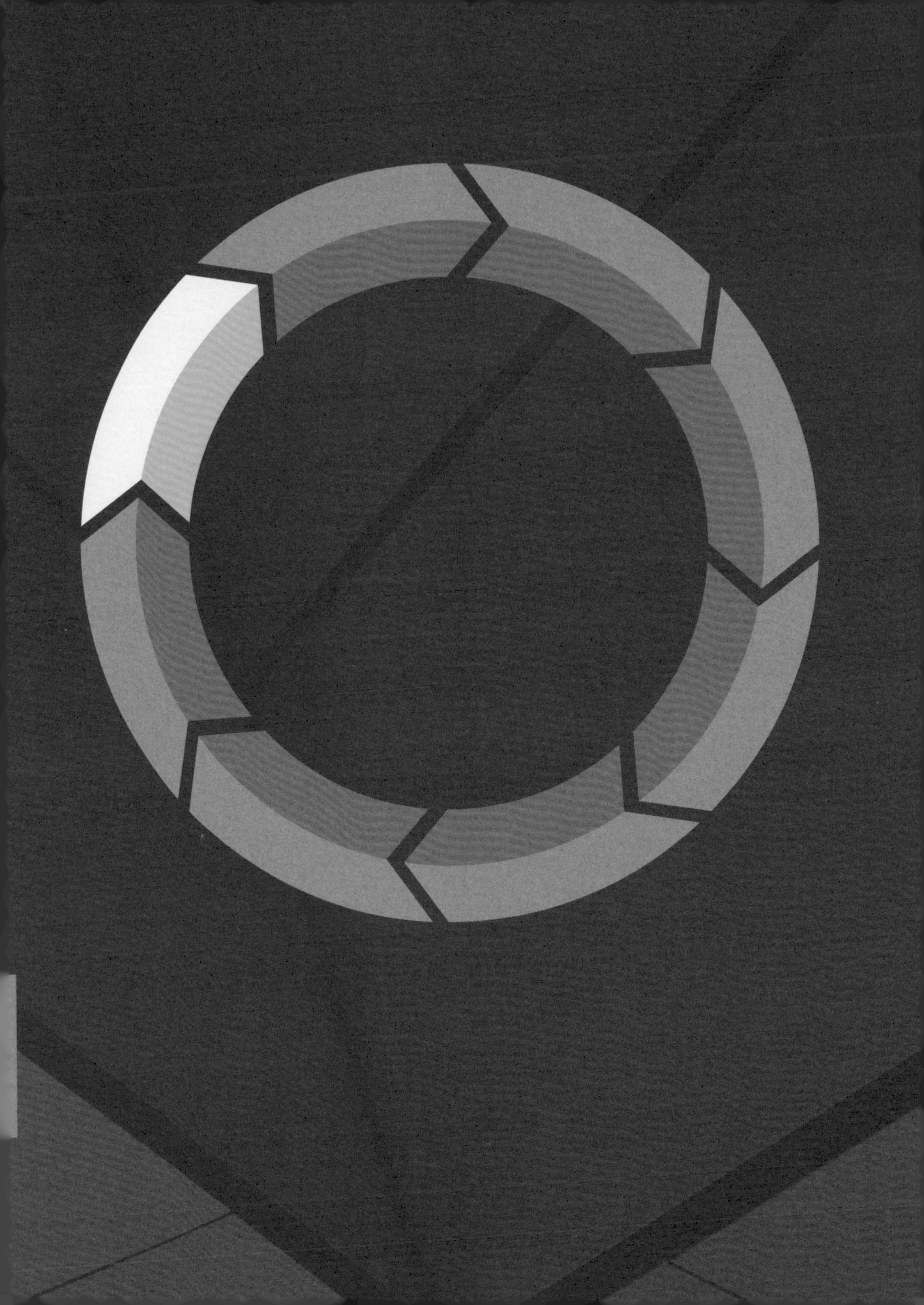

Stage 6

Handover and Close Out

Chapter overview

This chapter explains the role of the contract administrator during Stage 6 relating to the successful handover of the building and the conclusion of the Building Contract in line with the Project Programme. It considers the actions and procedures to be undertaken by the contract administrator in preparing for handover and during the rectification (defects liability) period. This will also include the inspection of any defects as they arise and are rectified and include methods and tools that can be used at the end of the rectification period through to the issue of the final certificate and the conclusion of the Building Contract.

The key coverage in this chapter is as follows:

What are the tasks and procedures before Practical Completion?

What are the tasks and procedures following Practical Completion?

What are the tasks and procedures at the end of the rectification period?

What are the tasks and procedures for the issue of the final certificate and concluding the Building Contract?

What are the liabilities for subsequent problems?

Why might the Project Information be archived?

What about Feedback?

Introduction

The priority of the project team, and particularly that of the contract administrator, during this stage will be to facilitate the successful handing over of the building by carrying out the activities listed in the Handover Strategy and, in the period following Practical Completion, concluding all aspects of the Building Contract. When considering the actions needed at Stage 6 the contract administrator should be aware that there is a degree of overlap with Stage 5, as can be seen in figure 5.1 in the previous chapter.

Although Practical Completion is achieved at the end of Stage 5, the actions required for Stage 6 start earlier, with the assembly of information and documentation needed to certify Practical Completion.

During Stage 6 work will be undertaken to prepare information, procedures and technical guidance prior to Practical Completion, so that the building is not only physically complete but is also ready for occupation when handed over to the client.

The remainder of Stage 6 of the RIBA Plan of Work 2013 immediately follows Practical Completion, when the client has taken possession of the building for occupation, fit-out or rental. The contract administrator will advise the client on the resolution of defects and have a monitoring role during the rectification period, responding to defects as they arise. Under the terms of the Building Contract, the contractor must rectify any defects if instructed to do so.

Practical Completion will have been achieved either as a single event or, if the Building Contract uses the sectional completion option, in a phased manner. The tasks and procedures in this chapter are based on a single completion date but, should the terms of the Building Contract require sectional completion, similar tasks and procedures should be undertaken for *each* section.

Unlike earlier stages, in which the type of procurement route has an impact on the tasks to be performed by the contract administrator, in Stage 6 the contract administrator's tasks will be the same for all forms of procurement. These will include:

- noting and inspecting defects
- issuing the certificate of making good defects
- agreeing the final account and issuing the final certificate.

However, in the case of the employer's agent in a design and build contract, any defects should be brought to the attention of the contractor as it is the contractor's responsibility to identify and rectify defects.

What are the Core Objectives of this stage?

The Core Objectives of the RIBA Plan of Work 2013 at Stage 6 are:

The Core Objectives at Stage 6 revolve around the handover of the completed building and the conclusion of the Building Contract. The Handover Strategy will have been prepared during Stage 1 and the contract administrator will refer to this during this stage and adjust the required activities accordingly.

What supporting tasks should be undertaken during Stage 6?

The Suggested Key Support Tasks in the RIBA Plan of Work 2013 have been devised to support the Core Objectives and to ensure that the documentation required to conclude the Building Contract has been prepared. One of these support tasks is carrying out activities related to the Handover Strategy, which has the most impact on the contract administrator. At this stage the Suggested Key Support Tasks for the contract administrator are:

- Carry out activities listed in Handover Strategy including Feedback for use during the future life of the building or on future projects.
- Updating of Project Information as required.

What is the purpose of the Handover Strategy?

The Handover Strategy reflects the increasing complexity of handing over a building at Practical Completion. Commissioning needs to be undertaken and users need to be trained in how to use the systems within the building (even residential projects use increasingly complex controls) or guided on how passive environmental measures should be used if they are to achieve the expected Project Outcomes. The increasing requirement on more complex projects for seasonal commissioning creates further challenges, with the growing recognition that not all aspects of the commissioning of a building and its engineering services systems can be carried out during the contract period. Seasonal commissioning recognises that some elements of the mechanical systems need to be commissioned when the external temperatures and indoor occupancy patterns are close to peak conditions for the time of year.

What are the tasks and procedures before Practical Completion

The handover of a project to the client at the end of the construction period is a very important stage of the project procurement process for ensuring the successful operation of the building. A well-organised, efficient and effective transfer of information to the client is essential. The contract administrator is integral to this process in that they have a direct relationship with the client and all members of the design team as well as with the main contractor. On larger buildings this would be undertaken in association with the facilities manager to assist in the management of the building and subsequent ongoing asset management.

As part of the Handover Strategy, the use of a 'soft landings' system can help to ensure the operational readiness of a building prior to Practical Completion. It will also make sure that the client receives support during the first few months of using the new building, and for up to three years after completion.

The term 'soft landings' refers to a strategy that seeks to ensure a seamless transition from construction to occupation and that operational performance is optimised. The Soft Landings Framework developed by BSRIA enables designers and contractors to improve the performance of buildings and generate Feedback for project teams. Since May 2011, soft landings principles have been in the Government Construction Strategy as a part of the government's effort to reduce the cost of public sector construction.

Where do I get more information on Soft Landings?

Free guidance and downloads on the Soft Landings Framework – the core principles, worksheets, etc. – are available from the BSRIA website at www.bsria.co.uk/services/design/soft-landings/

BSRIA Ltd, Old Bracknell Lane West, Bracknell, Berkshire, RG12 7AH

Tel: 01344 465600 Email: bsria@bsria.co.uk

What are the preparations for handover?

Common handover problems relate to inadequate preparation – where the building may be physically complete but it is not operationally ready. When the building is handed over to the client, it should be physically complete *and* ready for occupation.

To achieve this, the contract administrator should agree a sub-programme of work with the other consultants in good time, well ahead of commissioning, to ensure that the inspections are timely and not rushed.

It is not uncommon for the commissioning period to be squeezed as a result of delays that are outside the control of the contract administrator.

This sub-programme should include static commissioning, such as airtightness, soundproofing and other specific requirements, as well as all the building services commissioning, including renewable energy systems, ventilation, etc. In parallel there will be the requirement to prepare the 'As-constructed' Information, including operating and maintenance manuals, demonstrations, training and documentation, to ensure that the occupants are fully conversant with how the building works and how they are to use it from day one. Indeed, many problems at handover can be traced back to insufficient understanding by the occupier's staff of the technical systems which have been installed, particularly building services.

What documents are needed for handover?

Well ahead of handover the contract administrator should prepare a list of information that will be required prior to handover and from whom it needs to be obtained so that it can be assembled in good time for Practical Completion.

Pre-handover checklist/'As-constructed' Information

- Handover programme
- Training and familiarisation – allow adequate time
- Commissioning
- Drawing and specification details and/or BIM model
- Commissioning records
- Energy data
- Maintenance contracts
- Completion of building management system
- Meter readings at handover
- Operating and maintenance (O&M) manuals – the team should review the content with owners and tenants before they are signed off
- Health and safety file, completed by the health and safety adviser
- Technical guide, including clear descriptions in simple language on how the building's systems are intended to work (whereas the O&M manuals contain the technical details)
- Occupants' guide, to inform individuals about the systems and procedures – written for lay users
- Defects rectification process – for agreement

Arrange a pre-handover site meeting

The contract administrator should arrange a dedicated site meeting to discuss the handover process and to agree on requirements and outcomes. The meeting should be held at least four weeks prior to the proposed project completion date. However, on complex projects more time may be needed and so this meeting should be planned for much earlier, possibly by having an agenda item within the regular site progress meeting agenda thus allowing the pre-handover meeting to be more of a summary meeting.

The meeting should be attended by the client, design team members, the contractor and any relevant subcontractors and chaired by the contract administrator.

Agenda for pre-handover meeting

The agenda for a pre-handover meeting would include:

- Introduction and reason for the meeting
- Preparation of CAD information and/or BIM model
- Plant and equipment details
- Operating and maintenance manuals
- Health and safety features and measures
- Defects management and after-hours call-out procedures
- Connections and commissioning
- Systems operational training
- Approvals and certificates
- Warranties and guarantees
- Security systems (includes key handover process)
- Any other business

What is the health and safety file?

The health and safety file is a document required by the Construction (Design and Management) Regulations 2007 (CDM regulations). It should contain all the relevant health and safety information needed to allow the safe use of the building (including cleaning and maintenance) and ensure that any future construction work can be carried out safely. The file should be prepared by the health and safety adviser (CDM coordinator), and be completed in time for the building handover.

When is the building ready for handover?

The building will be physically complete at the end of Stage 5 but, unless there are exceptional circumstances, the certificate of Practical Completion should not be issued and the project handed over for occupation and use until the following six important 'contracted' activities have been completed:

- All systems, plant and equipment have been connected and commissioned, and all testing data and reports have been made available as part of the operating and maintenance manuals.
- All approvals, certificates, etc. have been issued.
- Management procedures for actions during the rectification period have been established.
- Demonstrations and training sessions on key systems and equipment have been successfully held to the satisfaction of the client.
- 'As-constructed' Information has been supplied, at least in draft form, prior to the project handover meeting.
- Call-out procedures have been agreed.

What happens with speculative buildings?

Usually, the building will be handed over as a completed building. However, on some projects, especially in the commercial sector, the building will be handed over incomplete, eg where the spaces are to be fitted-out by specialist teams.

In such instances it is essential that the fit-out team is made fully aware of the rules on what to do and what to avoid, otherwise serious problems could occur, particularly with heating and ventilating services. Indeed, it is preferable for the original design team to review the fit-out proposals if at all possible.

However, In speculative buildings this continuity is more difficult to achieve given the potential protracted timescales between Practical Completion and the arrival of the tenant. Nevertheless, even in such instances the client should be advised to retain the original design team to undertake a review when the tenant is known.

What are the tasks and procedures following Practical Completion?

Practical Completion at the end of Stage 5 triggers the start of the rectification period. At that point the contract administrator will have issued the Practical Completion certificate, issued the 'As-constructed' Information and advised the client of the actions to be taken as soon as they take over the building.

The contract administrator should also ensure that a copy of the health and safety file signed by the health and safety adviser is available together with the energy performance certificate and other documents for use during the rectification period.

As well as settling into the building, the client, with the assistance of any other users, will be identifying any minor issues that inevitably arise on occupation and which need to be attended to by the contractor.

What is the rectification period?

The rectification period – known as the defects liability period in previous versions of the RIBA Plan of Work – is the period during which the contractor is contractually obliged to rectify any defects that appear. The period starts immediately following Practical Completion and lasts for the time as stated in the Building Contract. This is usually six or 12 months – 12 months is advisable, even on small contracts, and particularly where a new heating system has been installed, to allow for a full seasonal cycle. The contract administrator has 14 days from the end of the rectification period to provide the contractor with a list of defects. For completeness, this should be a definitive list, including any items outstanding from Practical Completion as well as items which have arisen during the rectification period.

Remedying of initial defects

On the vast majority of projects the contractor will leave the site immediately following Practical Completion, unless there are any residual and minor defects left outstanding at Practical Completion to be completed. On a very large project, however, the contractor may, by agreement, arrange for a small team of operatives to remain on site during these early stages to rectify any defects as and when they arise.

In the likely event that there is no contractor present on site after Practical Completion then the contract administrator should suggest to the client that they compile a list of items needing attention as they arise. Non-urgent items should be highlighted for attention at the end of the rectification period. However, some defects will affect the operation of the building, such as the plumbing, heating and other systems, and will require urgent attention. In this instance the client should contact the contract administrator immediately to arrange for the contractor to carry out the work straight away. The non-urgent list can then be used to assist the contract administrator during their inspection(s) at the end of the rectification period.

The contract administrator should keep this record in a readily accessible format, such as that indicated in figure 6.1.

Typical form for defects reported after Practical Completion

Ed Associates Ltd.
141 New Business Park, Anytown, AT51 1ST

JOB No.	JOB TITLE			
123	**New Music Centre**			
	DEFECTS REPORTED AFTER PRACTICAL COMPLETION			Sheet 1 of 1
Date reported	location of defect	Description	Action by	Date completed
28-Aug-13	*Boiler Room*	*No Hot water*	*Plumbwell Ltd*	22-Sep-13
23-Oct-13	*Music Room 2*	*Door handle fallen off*	*B & D Ltd*	31-Oct-13
13-Dec-13	*Girls WC*	*extract fan not working*	*Electro Ltd*	15-Dec-13
12-Jan-14	*Staff Room*	*Crack under window**	*B & D Ltd*	
Complied by:	*Ed Smith*	* *items to be left until end of Rectification Period*	Date:	Jan-14

Figure 6.1 Form for defects reported after Practical Completion

What can be termed defects?

A defect is a physical problem in the building, whether in the fabric, structure or services, especially one that impairs correct function, that is not in accordance with the contract documents. The contract administrator owes a duty of care to the client to discover defects and bad workmanship. Defects can be grouped into four main categories:

- design deficiencies
- material deficiencies
- specification problems
- workmanship deficiencies.

When making inspections the contract administrator should identify items of work that are not in accordance with the Building Contract, such as where the contractor has ignored the drawings and specification and carried out the work in another way without the agreement of the contract administrator.

These are patent defects which may be open to view, exposed, manifest, evident or obvious, whereas the opposite, a latent defect, will exist before it is discovered as a hidden or concealed flaw in the work. When a latent defect becomes apparent, it ceases to be a latent defect and becomes patent.

Example of non-compliance

It is specified that an existing manhole which will be within a new extension is to be raised, but the contractor builds a precast floor over it and 'forgets' about it until the heating has been installed and the floor screed laid. As this is non-compliant with the Building Regulations and the building control officer will not approve it, the contract administrator has no alternative but to instruct the contractor to re-do the work correctly at the contractor's own cost.

Arrange for the release of half of the retention monies

Following the issuing of the Practical Completion certificate the contract administrator should arrange for the release of half of the retention monies. This requires the issue of a further interim certificate, either based on the contractor's valuation at Practical Completion or on the previous certificate if an updated valuation is not immediately available.

Note that on construction management contracts, a separate certificate of Practical Completion must be issued for each trade contract. This means that retention monies must be released as required for each individual

?

What are retention monies?

Retention is the withholding of payment to a contractor until the work is complete and free of defects and omissions. During the contract period, a percentage of the amount certified to the contractor on interim certificates (usually 5%) is retained by the client, although this will be subject to the wording in the Building Contract.

The contractor is usually given the right to the release of half of the retention monies at Practical Completion. The residual amount of retention is released by the issue of the final certificate, following the certification of making good defects.

trade contract. The same is true on management contracts, where each works contract must be certified individually.

Consider or review any extensions of time

If the contractor failed to complete the works by the agreed completion date the client has a remedy through the application of liquidated and ascertained damages. Although extensions of time can affect the client's claims for liquidated and ascertained damages, as they fix the actual construction period, the contractor's liability for liquidated and ascertained damages ends at Practical Completion.

The contract administrator has 12 weeks after Practical Completion to reassess any extensions of time awarded prior to Practical Completion, irrespective of whether the contractor has made further applications, and make any adjustments. This may include awarding more time as deemed necessary, in which case the contract administrator should then issue a further extension of time certificate.

Extensions of time are generally assessed and awarded during Stage 5 (see page 166), but a review at this stage enables the contract administrator to use, in addition to the other documents previously quoted, the 'as-built' Contract Programme as a reference point. This will indicate the actual time spent on each task and the overall impact of each 'relevant event' on the regular progress of the works.

The contract administrator should remember at all times to adopt a methodical approach to the award of extensions of time. There is a duty to be careful and to use a calculated approach, not an impressionistic one.

How are liquidated and ascertained damages applied?

Following the re-evaluation of the extensions of time, if it is found that the contractor still failed to complete within the extended period, the contract administrator should carry out the same tasks and procedures for the issue of a certificate of non-completion and the application of liquidated and ascertained damages as outlined in Stage 5 (see page 175).

How is financial information for the draft final account obtained?

Many contracts will stipulate the period allowed for the preparation of the final account, but much will depend on how long the contractor takes to compile the required information. Most contractors, especially smaller ones, will quickly move on to their next project, having received their payment after Practical Completion and half of the retention so that the final account is no longer their priority.

During the construction period and after Practical Completion, the contractor will be assembling their information for the final account. On large projects it is likely that this information will be provided by the contractor and others for passing to the cost consultant, who will be responsible for agreeing the final account. In this scenario the contract administrator will be acting as a conduit for the information between the contractor and the cost consultant. However, on smaller projects the final account would be prepared by the contractor and passed to the contract administrator for checking, comment and certification.

In either case, the cost consultant or the contract administrator will take into account all the variations and claims and will remeasure the works as necessary in order to prepare a draft final account. They will also take into account the individual figures for each of the sectional completion certificates.

What are the tasks and procedures at the end of the rectification period?

Towards the end of the rectification period the contract administrator should write to the client and contractor informing them of the need to

Flow chart for tasks and procedures at the end of the rectification period

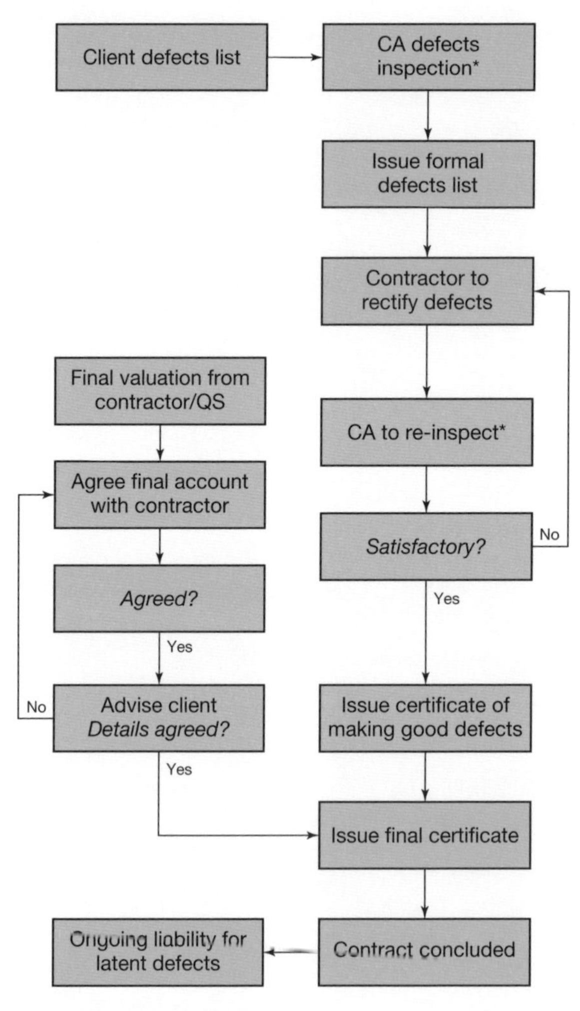

Figure 6.2 Flow chart for tasks and procedures at the end of the rectification period

inspect the building when the period expires. The contract administrator should liaise with the other consultants (where appointed) and arrange a suitable date for the inspection. Following the inspection, the contract administrator collates the information from consultants with their own and submits a schedule of defects to the contractor for action.

What about defects?

Towards the end of the rectification period the contract administrator should consult the client before the inspection to ascertain any items which the client has noticed which may not be immediately apparent on inspection. Most contractors are quite ready to meet their obligations and make good defects, but the contract administrator will occasionally have to deal with those who can require a lot of pressure before they put things right. Clients may also be unreasonable in their complaints, or may not want the work to be carried out during the rectification period or shortly afterwards, so compromises will be needed to ensure that the defects are completed in a timely manner and the contractor is paid within a reasonable timescale. Either way, all parties need to agree how the work will be done.

Defects on small projects

Some domestic clients may not want the contractor to carry out remedial work to avoid having their homes disturbed. In this instance, the client could be made a payment in lieu, so the client can arrange for the work to be done later. However, from experience, contractors are loath to do this as they are likely to have already expended that money, so any payment will effectively be a goodwill gesture.

All buildings will move and shrinkage cracks will inevitably occur during the rectification period, which the majority of contractors will put right. However, on small projects, builders, particularly smaller ones, will not repair minor cracks. It is therefore useful to know this in advance of the defects inspection. Alternatively, a clause could be inserted into the specification to say that such works should be allowed for.

The contract administrator will need to ensure that any serious remedial measures will be effective, such as to resolve water penetration, cracks etc., to avoid any possible latent defects.

How might the defects be recorded?

With the assistance of the list prepared by the client, inspections should be carried out systematically and recorded room by room and on an elevation-by-elevation basis. A standard pro-forma should be prepared to suit the practice requirement, with subsections for the various building elements, such as:

[*Room name*]

- Walls
- Floor
- Ceiling
- Internal doors
- External doors
- Windows
- Heating
- Plumbing
- Electrical
- Other

It is good practice to issue the list of defects no later than 14 days after the end of the rectification period.

What about re-inspecting the defects?

On notification from the contractor that all defects have been rectified a final inspection will be undertaken by the contract administrator and other consultants as necessary. This will follow the same format as the defects inspection, but will involve checking off the items on the defects list.

Once all of the defects have been rectified satisfactorily, a certificate of making good defects (such as the RIBA form in figure 6.3) appropriate to the form of contract being used should be signed and issued by the contract administrator. This is issued to the contractor and confirms that the contractor has completed the requirement to make good defects, shrinkages or other faults which were notified during or at the end of the rectification period.

Typical certificate of making good defects

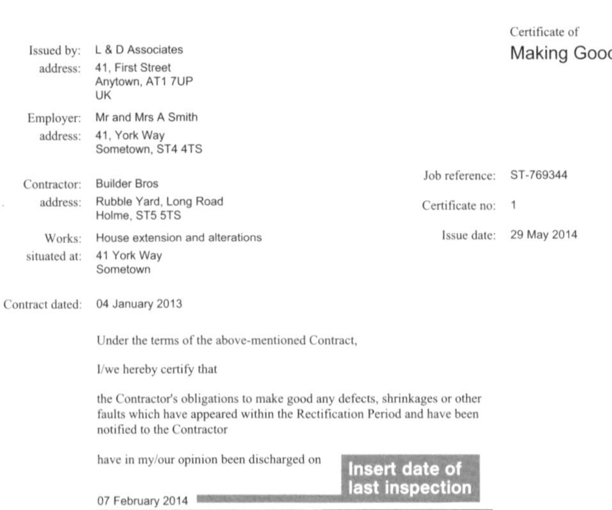

Certificate of
Making Good

Issued by:	L & D Associates
address:	41, First Street Anytown, AT1 7UP UK
Employer:	Mr and Mrs A Smith
address:	41, York Way Sometown, ST4 4TS
Contractor:	Builder Bros
address:	Rubble Yard, Long Road Holme, ST5 5TS
Works:	House extension and alterations
situated at:	41 York Way Sometown
Contract dated:	04 January 2013

Job reference:	ST-769344
Certificate no:	1
Issue date:	29 May 2014

Under the terms of the above-mentioned Contract,

I/we hereby certify that

the Contractor's obligations to make good any defects, shrinkages or other faults which have appeared within the Rectification Period and have been notified to the Contractor

have in my/our opinion been discharged on

07 February 2014

Insert date of last inspection

Sign

To be signed by or for the issuer named above

Signed ____________________

Distribution

☑ Employer [1]	☑ Structural Engineer [1]	☐ Planning Supervisor / CDM Coordinator	☐
☑ Contractor [1]	☐ M&E Consultant	☐	☐ Other
☐ Quantity Surveyor	☐ Clerk of Works	☐	☑ File [1]

for MW

Figure 6.3 Typical certificate of making good defects

What if the contractor delays or fails to complete the defects work?

The contractor has a contractual obligation to rectify any defects which are attributable to them if instructed to do so. However, it is not unknown for contractors, particularly small contractors, to lose interest in a project once they leave site, leaving the client with little option but to have the work done by others. The contract administrator should, in the first instance, contact the contractor personally to encourage them to complete their contractual obligation, advising that the fall-back option would be for the client to arrange to have the work done by others and claim the costs back from the contractor by deducting the cost from the outstanding retention monies.

In such circumstances clients may believe that they can arrange for the remedial work to be carried out to a higher quality using more expensive contractors, rather than the most cost-effective operatives. In this case the contract administrator should advise the client that under the Building Contract only the *amount that it would have cost the contractor to do the work can be claimed back*.

What are the tasks and procedures for the issue of the final certificate and concluding the Building Contract?

Following the receipt of the contractor's draft final account the contract administrator should check the figures and raise any queries prior to agreeing the final figure. When the final account is agreed, the contract administrator should, based on the final valuation, prepare and sign the final certificate.

Before the issue of the final certificate the contract administrator should obtain the client's agreement to the final account, having warned the client about potential 'extras'. However, neither the client nor the contractor should be allowed to influence the contract administrator's decision on the amount of the final sum due or when to issue the certificate.

The certificate (see figure 6.4) should be issued together with a copy of the agreed final account. This certificate should also release the remaining retention monies.

Typical final certificate

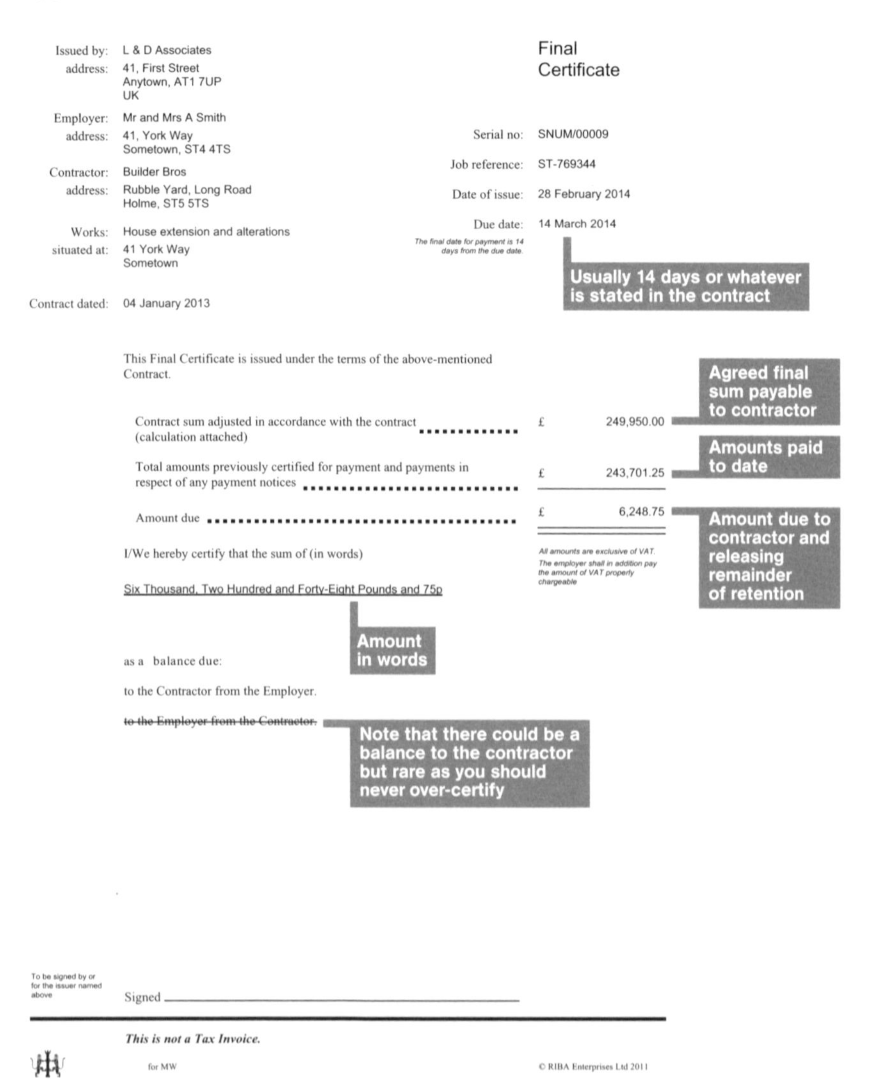

Issued by: L & D Associates
address: 41, First Street, Anytown, AT1 7UP, UK

Final Certificate

Employer: Mr and Mrs A Smith
address: 41, York Way, Sometown, ST4 4TS

Serial no: SNUM/00009

Contractor: Builder Bros
address: Rubble Yard, Long Road, Holme, ST5 5TS

Job reference: ST-769344

Date of issue: 28 February 2014

Works: House extension and alterations
situated at: 41 York Way, Sometown

Due date: 14 March 2014

The final date for payment is 14 days from the due date.

Contract dated: 04 January 2013

This Final Certificate is issued under the terms of the above-mentioned Contract.

Contract sum adjusted in accordance with the contract (calculation attached)	£	249,950.00
Total amounts previously certified for payment and payments in respect of any payment notices	£	243,701.25
Amount due	£	6,248.75

All amounts are exclusive of VAT. The employer shall in addition pay the amount of VAT properly chargeable

I/We hereby certify that the sum of (in words)

Six Thousand, Two Hundred and Forty-Eight Pounds and 75p

as a balance due:

to the Contractor from the Employer.

~~to the Employer from the Contractor.~~

To be signed by or for the issuer named above

Signed ____________________

This is not a Tax Invoice.

for MW

© RIBA Enterprises Ltd 2011

Figure 6.4 Typical final certificate

What is the impact of the final certificate?

The issue of the final certificate is conclusive as a statement of fact and law. Indeed, the contract administrator should not issue the final certificate until satisfied that the contract has been complied with. The contract administrator should ensure that the following matters have been concluded satisfactorily:

- The quality of materials, goods and workmanship reasonably reflect the requirements of contract documents.
- All extensions of time have been addressed.
- All adjustments have been made to the final contract sum.

Original certificates (such as the RIBA form in figure 6.4) appropriate to the form of contract being used should be issued to the contractor together with a breakdown of the final account, with copies sent to the client and the other consultants.

The final certificate should be issued within two months of the end of the rectification period or the issue of the certificate of making good defects, whichever is the later. Both the client and the contractor have the right to challenge the final certificate by commencing proceedings within 28 days of issue.

How might the client be notified that the Building Contract is concluded?

The client should be notified of the issue of the final certificate and advised of the implications of the ending of the Building Contract by letter, such as the example in figure 6.5.

How might the contractor be notified?

The contractor should be advised of the issue of the final certificate and advised of their future liabilities for latent defects by letter, such as in the example in figure 6.6.

Does the issue of the final certificate conclude the Building Contract?

The issue of the final certificate brings the authority of the contract administrator under the terms of the Building Contract to an end, although

Example letter to client issuing the final certificate

Dear XXXX,

Re: PROJECT TITLE

Further to our meeting on 26th February 2013 when it was confirmed that there were no outstanding defects, I now confirm the Final Account figure of £249,950.00 as before [OR WITH ADJUSTMENTS AS NECESSARY]. I therefore enclose the Final Certificate in the sum of £6,248.75 +VAT which releases the remaining retention monies together with a copy of the final valuation.

Therefore in accordance with the terms of the Contract, the sum of £7,498.50 should be paid to the Contractor within 14 days.

I enclose a Certificate of Making Good Defects for your records, but should there be any difficulties which could be termed latent defects, then please do not hesitate to contact me and I will contact the Contractor if contractually appropriate.

The release of this information now concludes the formal building contract and also brings us to the final fee stage for this part of the works. I will submit our Fee Account shortly which will be based on the final account figure.

I trust that all is in order and you will continue to enjoy using the new spaces which have been created.

Yours sincerely

Figure 6.5 Example letter to client issuing the final certificate

the contractor's liability (and indeed that of the consultants) continues until the end of the limitation period (see below).

What are the liabilities for subsequent problems?

Although the issue of the final certificate brings the contract to a close, the contractor and the design team still have ongoing liabilities for any unforeseen defects (latent defects). The parties will be liable for these until the end of the statutory limitation period.

Example letter to contractor issuing the final certificate

Dear XXXXX

Re: PROJECT TITLE

Following an inspection with the Client I am now able to confirm that the defects items have now been completed and on the basis that the Final Account is the same as the last valuation [AMEND AS NECESSARY] at Practical Completion in the sum of £249,950.00, I enclose the Final Certificate in the sum of £6,248.75+VAT which has been sent to the Employer for payment. This also releases the remaining retention monies.

In addition I also enclose the Certificate of Making Good Defects for your records.

The issue of the Final Certificate now concludes the formal building contract but should any subsequent matters arise which, in our opinion, could be termed latent defects then we will advise you accordingly.

It has been a pleasure to work with you on this project and I look forward to working with you again in the future.

Yours sincerely

Figure 6.6 Example letter to contractor issuing the final certificate

How does the limitation period affect liability?

The statutory limitation period is the period during which, according to law, the contractor and the project team may be liable for defects in the building.

Limitation periods are imposed by the Limitation Act 1980, which governs time limits for bringing different types of legal claims. The most relevant in construction are those for actions for breach of contract and tort, such as negligence, to be brought within a period of six years under a simple contract and 12 years if the contract is executed as a more formal deed. Unless otherwise noted in the contract, these time periods begin either on the date on which the breach of contract occurred, or the date the negligent act or omission occurred.

The Latent Damage Act 1986 introduced an extension to the ordinary six-year statutory limitation period by inserting section 14A into the Limitation

Act 1980. This extension is available for negligence claims for latent defects. This was intended to enable a claimant to have a reasonable period within which to conduct investigations, in particular by obtaining expert reports, but without leaving the period open ended.

Where there is a latent defect, the time limit is the later of six years from the date of accrual of the cause of action being raised, and three years from the earliest date on which the potential claimant knew, or reasonably ought to have known, material facts necessary to bring an action alleging negligence. This being subject to an overall limit of 15 years from the accrual of damage.

What is Post-occupancy Evaluation?

This has been defined within the RIBA Plan of Work 2013 as the evaluation undertaken after the occupation of the building to determine whether the Project Outcomes, both subjective and objective, as set out in the Final Project Brief and the Business Case have been achieved.

On a small project the level of service required will be minimal, but on a larger project there may be a client requirement to provide a post-project review of the design and construction process and an assessment of how the building fabric has performed against its predicted performance.

Why carry out these services?

On any project it is inevitable that initial problems will arise in the months following handover, which would be dealt with under the Building Contract at Stage 6. The principle of Post-occupancy Evaluations goes beyond whether the building has been built in accordance with the specification and drawings. Instead, it focuses on the building in use. Following handover the design team can assist the client in fine-tuning the building and its systems to ensure that, for example, sustainability targets are met. The emphasis is on obtaining a closer match between client and user expectations and the building as built by finding out what works well and what needs improving.

Everyone involved in the building, from end-users to facilities managers, through their experience of the building, is able to give valuable Feedback to help the building work more effectively. As a result, the client and occupiers will get the most out of the building in use.

What can be evaluated?

Irrespective of whether the BSRIA Soft Landings framework (see page 195) is used as the basis for the Stage 7 scope of works, the principles apply equally to any In Use evaluation.

The Post-occupancy Evaluation will include a review of Project Performance, Project Outcomes and updating the Project Information in response to ongoing Feedback. Feedback from users, facilities managers, maintenance staff, professionals and others can include responses to the following questions:

- Was there a major difference between the expectations and the Project Outcomes? For example, did the new prison reduce offending?
- How did the building fabric perform against the predicted performance?
- Was the building as energy efficient as predicted?
- Was the building easy to maintain?
- How did people use the building relative to the use perceived at the design stage?
- Was the building too complicated or difficult to look after?

Is there any input for the contract administrator?

Technically there is no role for the contract administrator at Stage 7 as their role is completed with the concluding of the contract at the end of Stage 6. However, given that there will inevitably be some overlap between Stages 6 and 7 the contract administrator may have some input to this stage, primarily in the updating of the Project Information and compiling of Feedback, from monitoring the services installation and so on.

Why might the Project Information be archived?

Once the Building Contract is concluded it is good practice for the members of the design team to archive their project records. This is not only important should any unforeseen matters arise in the future, but also it is good office housekeeping to leave space for new projects. Historically, with lots of drawing negatives and paper specifications, this was absolutely necessary, but even now, with mainly electronic documents, the recording and listing of the information to enable easy retrieval is of great importance.

Most of this information would have been prepared immediately prior to Practical Completion for inclusion in the handover documentation. At this stage the contract administrator should organise and collate all of the project documents in readiness for archiving. These should include hard copy or electronic versions of:

- contract documents
- selected working drawings
- BIM model
- specification
- certificates
- notes of inspections
- minutes
- letters to and from all parties
- emails.

While the project information should include a complete record of the documents used on the project, including those prepared by other members of the project team, it is not envisaged that the contract administrator would retain all of the documents (unless, perhaps, they are available in an electronic format). For example, if the contract administrator were also the architect, the practice would only archive their own drawings and specifications, leaving the other consultants to store theirs.

Once they have been compiled and catalogued the documents should be kept in safe and secure storage for a minimum of 15 years, with the start date commonly starting at the date for Practical Completion.

What about Feedback?

During the construction period the contract administrator will have gained a great deal of knowledge about the building as well as gaining Feedback for the practice. With this information the contract administrator would contribute to both in-house appraisals as well as any debriefing exercises with the client and others via Post-occupancy Evaluations.

It is important that the team obtain as much Feedback as possible to allow the practice to assess the performance of the in-house design team and other consultants appointed by the practice. Of even greater importance is the client's perception of the practice and reactions to the completed project. The purpose of this process is to assist the practice to improve performance and the standard of service provided.

How might the project be evaluated?

As soon as possible after the completion of the project a meeting should be held to appraise the following:

- the performance of the design team
- the working arrangements between the design team and the client
- the working arrangements between the design team and the contractor
- the contractor's performance
- the design solution, with particular reference to the client's comments
- the client's perception of the practice and the service provided
- reasons for variations, contract administrator's instructions and revisions to the drawings and design
- the quality system
- any other relevant matters.

The meeting should be chaired by the practice partners/directors, with the project architect, contract administrator and other staff attending as appropriate. Minutes should be taken and distributed to those required to take action. Larger projects are generally reviewed individually, whereas several smaller projects may be reviewed together.

It should be emphasised, however, that the meeting should not concentrate on negative aspects of the project but also identify the successes.

Observations and comments from the client and users (and general public where appropriate) should also be reviewed.

Should any specific complaints be made by the user or client on the performance of the design team, or on the performance of the building, materials or products, they should be recorded in a customer complaints file. Where ameliorative action is warranted, action should be taken at senior management level, including contacting the client on the matter.

Chapter summary 6

This chapter has demonstrated the tasks and procedures the contract administrator should complete to ensure the smooth handover of the building and the conclusion of the Building Contract.

Although other members of the team will be undertaking other roles during this stage it is the contract administrator who has a pivotal role in ensuring that the quality of the finished building is maintained and that all contractual matters are completed satisfactorily.

Throughout this process the contract administrator needs to be methodical in the procedures to be undertaken, accurate in evaluation and thorough in winding up the Building Contract. The residual information must be filed in such a way as to be easily retrievable and understandable so that, should there be a requirement to respond to a claim or litigation, the practice can evaluate the issues, particularly as the original personnel involved may no longer be employed.

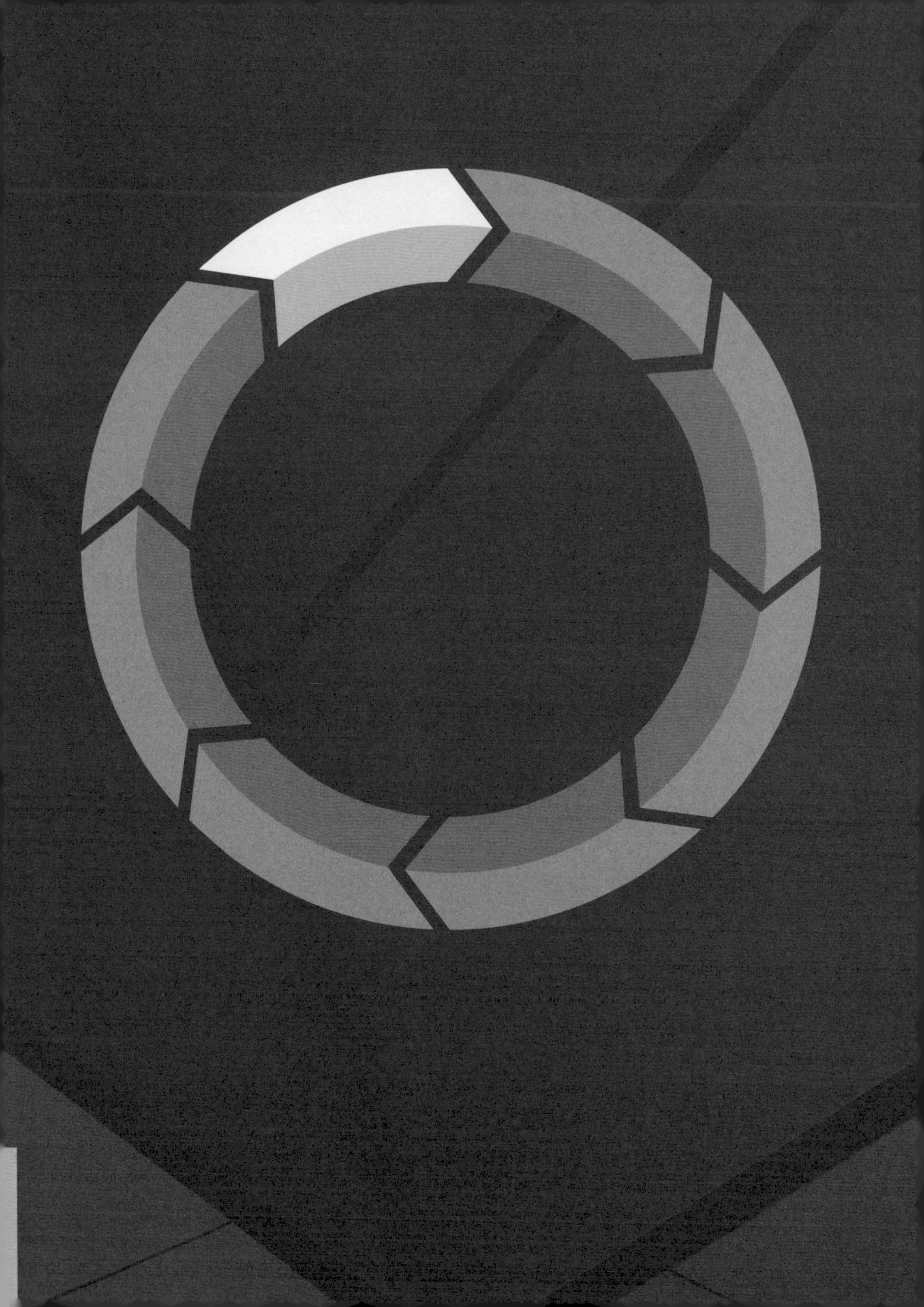

Stage 7

In Use

Chapter overview

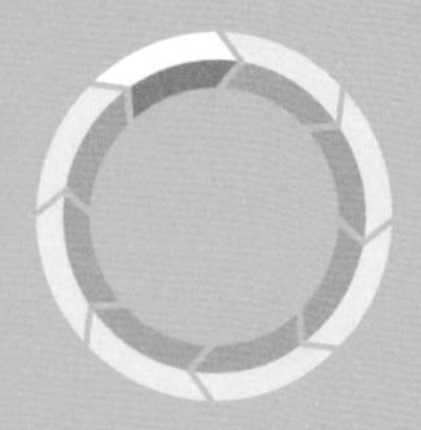

Stage 7 acknowledges the potential benefits of using the project design information to assist with the successful operation and use of the building. Many of the Stage 6 handover duties will be concurrent with this stage. Although the contract administrator's role will have been completed at the end of Stage 6, the contract administrator may have an input into evaluating the Project Outcomes based on their experience of administrating the Building Contract.

This chapter considers how the contract administrator role interfaces with other roles at this stage and how advice and decisions made at this stage are used as Feedback for future projects.

The key coverage in this chapter is as follows:

What is Post-occupancy Evaluation?

Feedback to Stage 0

What of the future?

Introduction

Stage 7 is a new stage within the RIBA Plan of Work 2013. It acknowledges the potential benefits of using the project design information to assist with the successful operation and use of a building throughout its life after handover and through to its eventual demolition. While the end of a building's life might be considered at Stage 7, it is more likely that Stage 0 of the follow-on project or refurbishment would deal with these aspects as part of strategically defining the future of the building or to inform other new projects.

Stage 7 includes Post-occupancy Evaluation and review of Project Performance as well as new duties that can be undertaken during the 'In Use' period of a building. It can also form part of the BSRIA Soft Landings framework, which was developed in response to the growing realisation that sustainability, energy efficiency and the overall performance of new and existing buildings needs to improve. Although the framework comes from a building services background, these principles are equally valid in other contexts whether or not the formal Soft Landings system is used.

It should be noted that Stage 7, although part of the core RIBA Plan of Work 2013 stages, might not be part of the scope of work for all or some of the design team. On larger, more complex projects, however, it should be seen as a defined work stage and so command an additional fee.

What are the Core Objectives of this stage?

The Core Objectives of the RIBA Plan of Work 2013 at Stage 7 are:

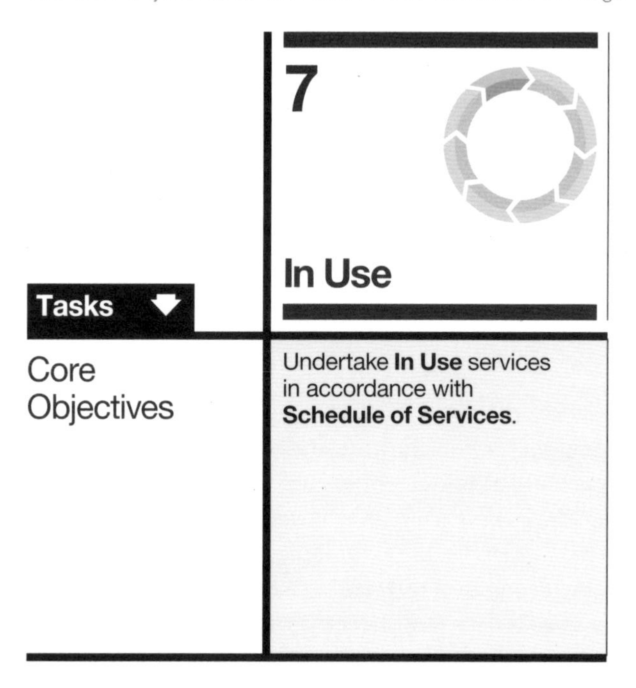

The Core Objectives at Stage 7 are wholly related to carrying out Post-occupancy Evaluations and other work (to be specifically agreed with the client on a job-by-job basis) so that relevant information can be used as Feedback to inform Stage 0 on subsequent projects. In general terms it is unlikely that the contract administrator will be directly involved at this stage, but their input from the construction stage will be invaluable.

What supporting tasks should be undertaken during Stage 7?

The Suggested Key Support Tasks noted in the RIBA Plan of Work 2013 have been devised to support the Core Objectives and to ensure that the documentation prepared will assist the user in managing their building better and to enable Feedback to future projects. At this stage the Suggested Key Support Tasks are:

- Conclude activities listed in Handover Strategy including Post-occupancy Evaluation, review of Project Performance, Project Outcomes and Research and Development aspects.
- Updating of Project Information, as required, in response to ongoing client Feedback until the end of the building's life.

The support tasks during this stage are focused on reviewing the building in use and updating Project Information as required, providing Feedback throughout the life of the building and informing future projects.

What is building performance evaluation?

There are two key ways of evaluating the building after handover: Post-occupancy Evaluation and building performance evaluation.

Post-occupancy Evaluation (see page 213) focuses on stakeholders and how the building is to operate and whether it supports their needs and aspirations, whereas building performance evaluation focuses on the performance of the building as a whole.

It is acknowledged that buildings often do not perform as well in use as the theoretical design, creating a shortfall in the performance of the building. Public and private clients alike have recognised this gap between predicted and actual performance and have begun to require operational performance targets within new-build contracts. In Stage 7 this underperformance can be analysed and understood to ultimately provide evidence to inform subsequent Stage 0 activity.

Building performance evaluation typically comprises study of the building as a set of interdependent systems relevant to the activities of the facilities management team. A frequent problem, especially in buildings with complex environmental requirements, is that system control interfaces are too complex for the operatives involved and, as a result, do not get used to full efficiency. Stage 7 can be invaluable for identifying this type of issue, allowing appropriate retrospective interventions to be developed and, most significantly, the avoidance of a repeat of the issue in subsequent projects.

Feedback to Stage 0

The appropriate information from the Post-occupancy Evaluation can be fed back to Stage 0, where data from the existing building can be used to inform the briefing process for the refurbishment and alteration of the building or even for similar new projects.

Feedback can be derived from materials produced as required by the Building Contract, such as the health and safety file, operating and maintenance manuals and drawings, and used to aid future objects. However, there is potential for Building Information Modelling (BIM) to have an increasing impact. The BIM model would have been used throughout the design and construction process and updated with as-built information at handover, in the same way as traditional drawings should be. The BIM model would be handed over to the client, where it should be updated and maintained as necessary therefore providing a good platform from which to launch future projects.

The other lessons from the Post-occupancy Evaluation, such as how people actually use the building, will similarly inform the development of the Strategic Brief for the next project.

What of the future?

The contract administrator needs to be alert to changes which will inevitably develop in the ongoing monitoring of a building in the future.

As greater emphasis is placed on the successful handover of the building at the end of Stage 5 there is a strong argument for increasing the defects period to three years. Indeed there has already been talk that the government is considering including three-year maintenance contracts as part of all public sector building projects. This would change the nature of the Building Contract and the tasks the contract administrator might have to undertake in this context.

Another potential area for change is Project Performance. Even buildings which were designed to be energy efficient have not always met their design targets when tested in use and, as a result, are using too much energy. Extending the contractor's liability to ensuring that the building meets the design criteria will bring the design standards and expectations into sharper focus, especially on larger projects.

Setting the Project Outcomes at an early stage with regard to energy and construction standards will focus the contractor on ensuring that target Project Outcomes are known, measured and met. This can be achieved by making the Maintenance and Operational Strategy more robust and by making the contractor responsible for ensuring that the building meets the design targets.

Chapter summary 7

Although the contract administrator will have completed their work at the end of Stage 6 and would not necessarily be involved at this stage, this chapter has looked at the principles of any Post-occupancy Evaluations.

By going through this exercise, the design team is able to identify good and bad experiences, which may benefit the future design and construction process. Furthermore, the people involved will take back something to their work, practice and the industry to improve future projects, whether related to the original building or not.

Contract administration glossary

Architect's instructions

see *Contract administrator's instructions.*

Bills of quantities

The traditional way of obtaining tenders for a project where the design is fully detailed at tender stage. Bills of quantities provide project-specific measured quantities for the items of work identified by the drawings and specifications in the tender documentation, allowing potential contractors to price the works on the same basis.

Building performance evaluation

A study of the building focusing on the performance of the building as a whole.

Collateral warranty

A contract that provides contractual rights to protect third parties who do not have a direct contractual relationship, enabling them to recover losses. They are used as a supporting document to a primary contract where an agreement needs to be put in place with a third party outside of the primary contract.

Compensation events

The terminology for variations, extensions of time and additional costs on NEC3 contracts.

Concurrent delay

A complex term, the legal status of which remains unclear. Delays on a project may have a number of causes, each of which will have a different contractual consequence.

Construction management

The Building Contract is administered and organised by a construction manager, but there is no direct contractual link between the construction manager and the trade contractors. There is a direct formal contract between the client and each individual trade contractor.

Contingencies

Sums within the Building Contract to cover unforeseen items and costs for things which might possibly happen. Not a catch-all for missed items!

Contract

A legally binding agreement which sets out the parties' responsibilities committing the parties to perform the terms of the contract and allocate risks.

Contract administrator

The party responsible for the administration of the Building Contract, including the issue of additional instructions and various certificates required to allow for handover and occupation of the building until all the defects have been rectified and the rectification period concluded.

Contract administrator's instructions

The means by which the contract administrator instructs the contractor. The Building Contract gives the contract administrator the authority to issue instructions, including making variations to the works, to remedy workmanship, goods or materials which are not in accordance with the contract.

Contract documents

These normally comprise three separate documents: the Building Contract, the specification and the drawings or plans.

Contract sum analysis

Prepared by a contractor as part of their tender on design and build projects. It breaks down the contractor's price into a format which allows the client and contract administrator to analyse it and compare it with the other tenders.

Contractor's designed portion

An agreement that the contractor designs specific parts of the works, particularly on JCT contracts. This should not be confused with design and build contracts where the contractor designs the whole of the works.

Daywork

A means by which a contractor is paid for specifically instructed work on the basis of the cost of labour, materials and plant plus a mark-up for overheads and profit when work cannot be priced in the normal way.

Defects

Materials, goods and workmanship that are not in accordance with the contract documents. Defects are a physical problem within the building, whether in the fabric, structure or services, especially one that impairs correct function.

Design and build procurement

Used for projects where the client employs a contractor to be responsible for both the design and construction of the building, acting as a single point of responsibility to the client.

Design team

Professionals typically appointed by the client to perform particular tasks on a project. The design team would be tailored to suit the particular needs of the project, and would commonly include the architect, health and safety adviser, cost consultant, services and structural engineers.

Electronic tendering

The exchange of tender documentation via portable media (eg DVDs and memory sticks), by email, through websites (extranets) and by using bespoke software.

Employer's agent

The agent acting on behalf of the client as the contract administrator for design and build contracts.

Exceptionally inclement weather

This term has no actual definition. In reality, any weather conditions may cause a delay, but only exceptionally adverse weather can give rise to an entitlement for an extension of time. It is up to the contract administrator to determine whether the weather was sufficiently adverse.

Extensions of time

Extensions to the construction period, allowed where the contractor is delayed by an event that is not their fault under the terms of the Building Contract.

FIDIC

The International Federation of Consulting Engineers (FIDIC is the acronym for the French name: Fédération Internationale des Ingénieurs-Conseil), an international standards organisation for the construction industry. FIDIC contracts are commonly used where tenders are invited on an international basis, particularly in countries where standard forms do not exist or are unsuitable.

Final certificates

Confirmation by the contract administrator that the Building Contract has been fully completed. The final certificate is issued at the end of the rectification period and has the effect of releasing all remaining monies due to the contractor.

Force majeure

Events beyond the control of the contracting parties, commonly described as 'acts of God', which prevent or impede the contractor's obligations under the contract. Such events may entitle either party to suspend or terminate the contract.

Interim certificates

The mechanism to initiate payments in instalments by the client to the contractor before the works are complete.

Joint Contracts Tribunal (JCT)

The JCT has produced standard forms of contract, guidance notes and other standard documentation for use in the construction industry since it was established in 1931 by the RIBA and the National Federation of Building Trades Employers (NFBTE).

Latent defects

Defects that exist as hidden or concealed flaws in the work. When a latent defect becomes apparent it ceases to be a latent defect and becomes patent.

Letter of intent

An expression of an intention to enter into a contract at a future date without creating a contractual relationship until the future contract has been entered into. It is not an 'agreement to agree'. The term does not have an enforceable legal meaning, such as subject to contract.

Limitation periods

Time limits within which a party to a contract can bring a claim. Limitation periods are imposed by statute under the Limitation Act 1980.

Loss and expense

Provision in construction contracts for the contractor to claim direct loss and/or expense as a result of the progress of the works being materially affected by events for which the client is responsible.

Management contracting

This form of procurement relies on the management skills of the contractor and is suitable for large projects with complex requirements. The management contractor is appointed at the initial stages of a project. Management contracts rely on teamwork and trust and provide the maximum overlap of design, procurement and construction.

NEC

The NEC is a family of standard contracts. Originally launched in 1993 as the New Engineering Contract, it aims to promote a collaborative and integrated working approach.

Novation

A process by which contractual rights and obligations are transferred from one party to another. On a building project, the novation occurs when design consultants who were initially contracted to the client then start working for the contractor.

One-stage tendering

The most common method of inviting tenders, whereby the contractor submits their proposals and price and the contract is let on the basis of that tender.

Partial possession

At completion, the client or tenant may take possession of part of the building even if the works are ongoing or there are defects that have not been rectified and sectional completion has not been specified.

Partnering or collaborative working

A broad term used to describe a collaborative management approach that encourages openness and trust between parties to a contract. The parties become dependent on one another for success. This requires a change in culture, attitude and procedures throughout the supply chain. It is most commonly used on large, long-term or high-risk contracts.

PAS 91:2013

A standardised pre-qualification questionnaire sponsored by the Department for Business, Innovation and Skills.

Patent defects

Defects which may be open to view, exposed, manifest, evident or obvious.

Periodic inspections

Periodic visits to the site to visually monitor the progress of the works. Inspection should not normally involve detailed checking of dimensions or testing of materials, and should not be confused with supervision of the work on a day-to-day basis by the contractor.

Preliminaries and general conditions

Tender documents that describe the works as a whole in a clear and precise way, and specify general conditions and requirements, such as contract details, time constraints, approvals, testing and completion.

Pre-qualification

A means of proving the capabilities of potential contractors before they are invited to tender. A list of potential contractors is prepared, any one of which could do the job. This helps to simplify the process of choosing the contractor after the receipt of tenders.

Pre-qualification questionnaires

Questionnaires that test potential contractors against basic eligibility criteria. They are intended to simplify the process of selecting the contractor and reduce the likelihood of non-compliant tenders being submitted.

Procurement

A generic term embracing all those activities undertaken by a client seeking to bring about the design and construction of a new-build project or the refurbishment of a building.

Project team

Comprises the design team, the client and the contractor. How the members are connected will depend on the procurement route and how the contractor was introduced to other members of the design team.

Provisional sums

Allowances included in tenders to cover specific elements of the works which are not yet defined enough for the contractors to price.

Quasi-arbitrator

Someone, usually the contract administrator, who is empowered under the Building Contract to resolve disputes between the main contracting parties. As this person is usually appointed by the client but has a duty to be fair to both parties, the role can be deemed to be 'quasi' (def: seemingly; apparently but not really).

Rectification period

Formerly known as the defects liability period, a period of usually six or 12 months after Practical Completion during which the contractor is obliged to make good any defects which are attributable to them.

Retention monies

Sums of money that are withheld from a contractor until the works have been completed and are free of defects and omissions.

Schedules of rates

Used where insufficient detail is available at the time of tender. In its simplest form a schedule of rates can be a list of staff, types of labour and plant hire rates and rates to cover all likely activities that might form part of the works.

Schedules of work

Lists of the work required, arranged on an elemental basis (eg brickwork, plastering, decorations) or on a room-by-room basis. A schedule that includes a description of the work required is known as a specified schedule of work.

Sectional completion

A provision within construction contracts allowing different completion dates to be set for different sections of the works.

Snagging

This term does not have an agreed meaning: it is a slang expression, not a contractual term. It is widely used to define the process of inspection necessary to compile a list of minor defects or omissions in building works for the contractor to rectify.

Soft Landings

A BSRIA strategy to ensure that the transition from construction to occupation is seamless and that operational performance is optimised.

Specifications

Documents which define the products to be used, the quality of work and any performance requirements or conditions under which the work is to be executed. Specifications may be prescriptive or performance based.

Tendering

A method for selecting the contractor to carry out the contract works. Compliant bids are invited from interested contractors and evaluated in a fair and transparent way.

Third party rights

The right of persons who are not a party to a contract (the third party) to enforce the benefit of a term of that contract, such as a funder enforcing a term of a Building Contract between the client and a professional or the contractor.

Time at large

A situation where no date has been set for completion, or where the date for completion has become invalid, thus the contractor is no longer bound by the obligation to complete the works by the date set.

Traditional procurement

Historically the most common form of procurement for all sizes and types of building project. It gives a clear distinction between the design and construction stages, whereby the design team prepares the detailed design and the contractor builds to the information they have been given.

Two-stage tendering

The preferred contractor is selected initially on the basis of the outcome of the first-stage competitive tender, based on pricing the tender documents using preliminary design information. The contractor will then be given a limited appointment to start work on collaborating with the design team in the development of the design, leading to the compilation of the final price after the second stage.

Variations

Alterations to the scope of works in a Building Contract, in the form of additions, substitutions or omissions from the original scope of works.

RIBA Plan of Work 2013 glossary

A number of new themes and subject matters have been included in the RIBA Plan of Work 2013. The following presents a glossary of all of the capitalised terms that are used throughout the RIBA Plan of Work 2013. Defining certain terms has been necessary to clarify the intent of a term, to provide additional insight into the purpose of certain terms and to ensure consistency in the interpretation of the RIBA Plan of Work 2013.

'As-constructed' Information

Information produced at the end of a project to represent what has been constructed. This will comprise a mixture of 'as-built' information from specialist subcontractors and the 'final construction issue' from design team members. Clients may also wish to undertake 'as-built' surveys using new surveying technologies to bring a further degree of accuracy to this information.

Building Contract

The contract between the client and the contractor for the construction of the project. In some instances, the **Building Contract** may contain design duties for specialist subcontractors and/or design team members. On some projects, more than one Building Contract may be required; for example, one for shell and core works and another for furniture, fitting and equipment aspects.

Building Information Modelling (BIM)

BIM is widely used as the acronym for 'Building Information Modelling', which is commonly defined (using the Construction Project Information Committee (CPIC) definition) as: 'digital representation of physical and functional characteristics of a facility creating a shared knowledge resource for information about it and forming a reliable basis for decisions during its life cycle, from earliest conception to demolition'.

Business Case

The **Business Case** for a project is the rationale behind the initiation of a new building project. It may consist solely of a reasoned argument. It may contain supporting information, financial appraisals or other background information. It should also highlight initial considerations for the **Project Outcomes**. In summary, it is a combination of objective and subjective considerations. The **Business Case** might be prepared in relation to, for example, appraising a number of sites or in relation to assessing a refurbishment against a new build option.

Change Control Procedures

Procedures for controlling changes to the design and construction following the sign-off of the Stage 2 Concept Design and the **Final Project Brief**.

Common Standards

Publicly available standards frequently used to define project and design management processes in relation to the briefing, designing, constructing, maintaining, operating and use of a building.

Communication Strategy

The strategy that sets out when the project team will meet, how they will communicate effectively and the protocols for issuing information between the various parties, both informally and at Information Exchanges.

Construction Programme

The period in the **Project Programme** and the **Building Contract** for the construction of the project, commencing on the site mobilisation date and ending at **Practical Completion**.

Construction Strategy

A strategy that considers specific aspects of the design that may affect the buildability or logistics of constructing a project, or may affect health and safety aspects. The **Construction Strategy** comprises items such as cranage, site access and accommodation locations, reviews of the supply chain and sources of materials, and specific buildability items, such as the choice of frame (steel or concrete) or the installation of larger items of plant. On a smaller project, the strategy may be restricted to the location of site cabins and storage, and the ability to transport materials up an existing staircase.

Contractor's Proposals

Proposals presented by a contractor to the client in response to a tender that includes the **Employer's Requirements**. The **Contractor's Proposals** may match the **Employer's Requirements**, although certain aspects may be varied based on value engineered solutions and additional information may be submitted to clarify what is included in the tender. The **Contractor's Proposals** form an integral component of the **Building Contract** documentation.

Contractual Tree

A diagram that clarifies the contractual relationship between the client and the parties undertaking the roles required on a project.

Cost Information

All of the project costs, including the cost estimate and life cycle costs where required.

Design Programme

A programme setting out the strategic dates in relation to the design process. It is aligned with the **Project Programme** but is strategic in its nature, due to the iterative nature of the design process, particularly in the early stages.

Design Queries

Queries relating to the design arising from the site, typically managed using a contractor's in-house request for information (RFI) or technical query (TQ) process.

Design Responsibility Matrix

A matrix that sets out who is responsible for designing each aspect of the project and when. This document sets out the extent of any performance specified design. The **Design Responsibility Matrix** is created at a strategic level at Stage 1 and fine tuned in response to the Concept Design at the end of Stage 2 in order to ensure that there are no design responsibility ambiguities at Stages 3, 4 and 5.

Employer's Requirements

Proposals prepared by design team members. The level of detail will depend on the stage at which the tender is issued to the contractor. The **Employer's Requirements** may comprise a mixture of prescriptive elements and descriptive elements to allow the contractor a degree of flexibility in determining the **Contractor's Proposals**.

Feasibility Studies

Studies undertaken on a given site to test the feasibility of the **Initial Project Brief** on a specific site or in a specific context and to consider how site-wide issues will be addressed.

Feedback

Feedback from the project team, including the end users, following completion of a building.

Final Project Brief

The **Initial Project Brief** amended so that it is aligned with the Concept Design and any briefing decisions made during Stage 2. (Both the Concept Design and **Initial Project Brief** are Information Exchanges at the end of Stage 2.)

Handover Strategy

The strategy for handing over a building, including the requirements for phased handovers, commissioning, training of staff or other factors crucial to the successful occupation of a building. On some projects, the Building Services Research and Information Association (BSRIA) Soft Landings process is used as the basis for formulating the strategy and undertaking a **Post-occupancy Evaluation** (www.bsria.co.uk/services/design/soft-landings/).

Health and Safety Strategy

The strategy covering all aspects of health and safety on the project, outlining legislative requirements as well as other project initiatives, including the **Maintenance and Operational Strategy**.

Information Exchange

The formal issue of information for review and sign-off by the client at key stages of the project. The project team may also have additional formal **Information Exchanges** as well as the many informal exchanges that occur during the iterative design process.

Initial Project Brief

The brief prepared following discussions with the client to ascertain the **Project Objectives**, the client's **Business Case** and, in certain instances, in response to site **Feasibility Studies**.

Maintenance and Operational Strategy

The strategy for the maintenance and operation of a building, including details of any specific plant required to replace components.

Post-occupancy Evaluation

Evaluation undertaken post occupancy to determine whether the **Project Outcomes**, both subjective and objective, set out in the **Final Project Brief** have been achieved.

Practical Completion

Practical Completion is a contractual term used in the **Building Contract** to signify the date on which a project is handed over to the client. The date triggers a number of contractual mechanisms.

Project Budget

The client's budget for the project, which may include the construction cost as well as the cost of certain items required post completion and during the project's operational use.

Project Execution Plan

The **Project Execution Plan** is produced in collaboration between the project lead and lead designer, with contributions from other designers and members of the project team. The **Project Execution Plan** sets out the processes and protocols to be used to develop the design. It is sometimes referred to as a project quality plan.

Project Information

Information, including models, documents, specifications, schedules and spreadsheets, issued between parties during each stage and in formal Information Exchanges at the end of each stage.

Project Objectives

The client's key objectives as set out in the **Initial Project Brief**. The document includes, where appropriate, the employer's **Business Case**, **Sustainability Aspirations** or other aspects that may influence the preparation of the brief and, in turn, the Concept Design stage. For example, **Feasibility Studies** may be required in order to test the **Initial Project Brief** against a given site, allowing certain high-level briefing issues to be considered before design work commences in earnest.

Project Outcomes

The desired outcomes for the project (for example, in the case of a hospital this might be a reduction in recovery times). The outcomes may include operational aspects and a mixture of subjective and objective criteria.

Project Performance

The performance of the project, determined using **Feedback**, including about the performance of the project team and the performance of the building against the desired **Project Outcomes**.

Project Programme

The overall period for the briefing, design, construction and post-completion activities of a project.

Project Roles Table

A table that sets out the roles required on a project as well as defining the stages during which those roles are required and the parties responsible for carrying out the roles.

Project Strategies

The strategies developed in parallel with the Concept Design to support the design and, in certain instances, to respond to the **Final Project Brief** as it is concluded. These strategies typically include:

- acoustic strategy
- fire engineering strategy
- **Maintenance and Operational Strategy**
- **Sustainability Strategy**
- building control strategy
- **Technology Strategy.**

These strategies are usually prepared in outline at Stage 2 and in detail at Stage 3, with the recommendations absorbed into the Stage 4 outputs and Information Exchanges.

The strategies are not typically used for construction purposes because they may contain recommendations or information that contradict the drawn information. The intention is that they should be transferred into the various models or drawn information.

Quality Objectives

The objectives that set out the quality aspects of a project. The objectives may comprise both subjective and objective aspects, although subjective aspects may be subject to a design quality indicator (DQI) benchmark review during the **Feedback** period.

Research and Development

Project-specific research and development responding to the **Initial Project Brief** or in response to the Concept Design as it is developed.

Risk Assessment

The **Risk Assessment** considers the various design and other risks on a project and how each risk will be managed and the party responsible for managing each risk.

Schedule of Services

A list of specific services and tasks to be undertaken by a party involved in the project which is incorporated into their professional services contract.

Site Information

Specific **Project Information** in the form of specialist surveys or reports relating to the project- or site-specific context.

Strategic Brief

The brief prepared to enable the Strategic Definition of the project. Strategic considerations might include considering different sites, whether to extend, refurbish or build new and the key **Project Outcomes** as well as initial considerations for the **Project Programme** and assembling the project team.

Sustainability Aspirations

The client's aspirations for sustainability, which may include additional objectives, measures or specific levels of performance in relation to international standards, as well as details of specific demands in relation to operational or facilities management issues.

The **Sustainability Strategy** will be prepared in response to the **Sustainability Aspirations** and will include specific additional items, such as an energy plan and ecology plan and the design life of the building, as appropriate.

Sustainability Strategy

The strategy for delivering the **Sustainability Aspirations**.

Technology Strategy

The strategy established at the outset of a project that sets out technologies, including Building Information Modelling (BIM) and any supporting processes, and the specific software packages that each member of the project team will use. Any interoperability issues can then be addressed before the design phases commence.

This strategy also considers how information is to be communicated (by email, file transfer protocol (FTP) site or using a managed third party common data environment) as well as the file formats in which information will be provided. The **Project Execution Plan** records agreements made.

Work in Progress

Work in Progress is ongoing design work that is issued between designers to facilitate the iterative coordination of each designer's output. Work issued as **Work in Progress** is signed off by the internal design processes of each designer and is checked and coordinated by the lead designer.

Index

Note: page numbers in italics refer to figures; page numbers in bold refer to tables.